Synthesis Lectures on Sustainable Development

Series Editor

Thomas Siller, Civil and Environmental Engineering, Colorado State University, Fort Collins, USA

This series publishes short books related to sustainable development practices relevant to engineers, technologists, managers, educators, and policy makers. The books are organized around the United Nations Sustainable Development Goals for 2015–2030. Design for sustainability along with the economics of sustainable development will be the common themes for each book. Topics to be covered will span all the major engineering disciplines along with the natural and environmental sciences that contribute to an understanding of sustainable development. The goal of this series is to make theory and research accessible to practitioners working on sustainable development efforts.

Atul H. Bari · Hemantkumar N. Akolkar ·
Navnath T. Hatvate · A. K. Haghi

Wastewater Treatment

An Introduction to Concepts and Practice

Springer

Atul H. Bari
Institute of Chemical Technology
Mumbai, Marathwada Campus
Jalna, Maharashtra, India

Hemantkumar N. Akolkar
Department of Chemistry
Abasaheb Marathe College
Ahmednagar, Maharashtra, India

Navnath T. Hatvate
Institute of Chemical Technology
Mumbai, Marathwada Campus
Jalna, Maharashtra, India

A. K. Haghi
Department of Chemistry
Institute of Molecular Sciences
University of Coimbra
Coimbra, Portugal

ISSN 2637-7675 ISSN 2637-7691 (electronic)
Synthesis Lectures on Sustainable Development
ISBN 978-3-032-07584-0 ISBN 978-3-032-07585-7 (eBook)
https://doi.org/10.1007/978-3-032-07585-7

This Springer imprint is published by the registered company Springer Nature Switzerland AG
The registered company address is: Gewerbestrasse 11, 6330 Cham, Switzerland

Preface

Modern environmental management must include wastewater treatment since it is essential to maintaining the quality of our water supplies and safeguarding public health. There has never been a greater need for efficient wastewater treatment systems due to the world's population growth and urbanization.

This book *Wastewater Treatment—An Introduction to Concepts and Practice* covers all the steps and procedures involved in treating wastewater, from basic treatment to sludge disposal, and offers a thorough introduction to the principles and practices of wastewater treatment. The book offers a clear and succinct summary of the main ideas and technologies in wastewater treatment, making it an invaluable tool for academics, practitioners, and students.

The book is organized into five chapters, each focusing on a specific stage of wastewater treatment. Chapter 1 introduces the basic concepts of wastewater treatments. Chapter 2 delves into preliminary treatment stage, covering topics such as screening, grit removal, and large object removal primary treatment and primary treatment techniques including primary sedimentation, primary clarifiers, and primary sludge treatment. Chapter 3 explores secondary treatment, discussing the activated sludge process, membrane bioreactors, and disinfection. Chapter 4 examines tertiary treatment, including sand filtration, membrane filtration, and various advanced oxidation processes. Finally, Chap. 5 addresses sludge treatment and disposal, covering topics such as thickening, dewatering, stabilization, and sludge reuse.

We have focused on the practical aspects of wastewater treatment throughout the book, emphasizing the importance of effective treatment system design, operation, and maintenance. It is our aim that this book will be a helpful resource for wastewater treatment professionals and help create sustainable and efficient wastewater management solutions.

We hope to give readers the information and abilities they need to handle the intricate problems related to wastewater management by giving them a comprehensive grasp of the theories and procedures of wastewater treatment. Everyone with an interest in wastewater treatment, from researchers and students to practitioners and policymakers, will find this book to be a useful resource.

Jalna, India Dr. Atul H. Bari
Ahmednagar, India Dr. Hemantkumar N. Akolkar
Jalna, India Dr. Navnath T. Hatvate
Edinburgh, UK Dr. A. K. Haghi

Contents

About the Authors

Dr. Atul H. Bari Ph.D. is an assistant professor at the Institute of Chemical Technology, Mumbai, Marathwada Campus, Jalna, Maharashtra, India. He holds a Masters and a Ph.D. in of Chemical Engineering from the Department of Chemical Engineering at the Institute of Chemical Technology, Mumbai, Maharashtra, India. He has 15 years of research and five years of teaching experience. His areas of expertise and research interests include the Mathematical modelling and simulations, Population balance modelling for particulate processes, wastewater treatment, process intensification, crystallization, nanoparticle synthesis using flow synthesis, chemical kinetics. Dr. Bari has published over 15 research publications in international journals.

Hemantkumar N. Akolkar Ph.D. is an Associate Professor in the Department of Chemistry at Rayat Shikshan Sanstha's Abasaheb Marathe Arts and New Commerce Science College, Maharashtra, India. He is an academician and researcher who has taught for the last 15 years at both undergraduate and postgraduate levels. He has published over 65 research papers in national and international journals and conferences of repute. Also, he has published six Indian patents and seven research-oriented books with international publishers like Springer, Royal Society of Chemistry, IGI Global etc. His areas of interest are heterocyclic chemistry, synthetic organic chemistry, and green chemistry. He is a reviewer of several journals of international repute in chemistry.

Dr. Navnath T. Hatvate Ph.D. is an assistant professor at the Institute of Chemical Technology, Mumbai, Marathwada Campus, Jalna. He holds a Master of Pharmacy and a Ph.D. from the Department of Pharmaceutical Sciences and Technology at the Institute of Chemical Technology, Mumbai. He has 12 years of research and six years of teaching experience. His areas of expertise and research interests include the development of new chemical entities (NCE) and their analogues, modifications of excipients and their applications in drug delivery, the use of AI and ML in drug discovery and drug delivery, process chemistry of APIs and intermediates, and total synthesis of natural products, bio-active compounds, application of novel technologies in organic synthesis. Dr. Hatvate has published over 50 research publications and authored four books in international journals. He serves as an editorial board member for Scientific Reports and Discover Pharmaceutical Sciences. (Nature Publishing).

A. K. Haghi Ph.D. is a retired professor and has written, co-written, edited, or co-edited more than 1000 publications, including books, book chapters, and papers in refereed journals with over 4500 citations and H-index of 35, according to the Google Scholar database. Professor Haghi holds a B.Sc. in urban and environmental engineering from the University of North Carolina (USA) and has two M.Sc. degrees, one in mechanical engineering from North Carolina State University (USA) and another in applied mechanics, acoustics, and materials from the Université de Technologie de Compiègne (France). He was awarded a Ph.D. in engineering sciences at Université de Franche-Comté (France). He is a regular reviewer of leading international journals. He is appointed Honorary Research Associate (HRA) at the University of Coimbra, Portugal.

Introduction to Wastewater Treatment 1

Abstract

The growing demand for clean water and rapid industrialization have highlighted the urgent need for effective wastewater treatment. Untreated or poorly managed wastewater poses significant threats to public health, aquatic ecosystems, and environmental sustainability. This chapter provides a comprehensive introduction to wastewater treatment, beginning with its importance in reducing pollution and conserving water resources. It defines key terminologies essential for understanding treatment processes, establishing a solid foundation for students, researchers, and professionals new to the field. The chapter also examines various wastewater sources, including domestic, industrial, and agricultural origins, and outlines their unique chemical and physical characteristics. Finally, the chapter offers an overview of both conventional and emerging treatment technologies, emphasizing the objectives of each stage and their roles in achieving safe and sustainable water reuse. This foundational chapter sets the stage for a deeper exploration of advanced treatment methods discussed in subsequent sections.

1.1 The Need for Wastewater Treatment

Water contamination due to unchecked discharge of industrial and municipal effluents has long posed serious environmental and public health challenges. These effluents introduce a range of organic and inorganic pollutants into natural water bodies, with heavy metals being hazardous due to their toxicity, persistence, and bioaccumulative nature. Once consumed by aquatic organisms, such pollutants can move up the food chain, eventually affecting human health. Health impacts include teratogenicity, carcinogenicity, organ damage, and neurological impairment, emphasizing the urgent need for preventive wastewater treatment measures [1, 2]. Beyond heavy metals, a wide variety of contaminants-including

A. H. Bari et al., *Wastewater Treatment*, Synthesis Lectures on Sustainable Development,
https://doi.org/10.1007/978-3-032-07585-7_1

pharmaceuticals, synthetic dyes, pesticides, and pathogens-have been documented in natural waters. These compounds are increasingly resistant to degradation and are known to cause chronic toxicity, reproductive issues, and immune suppression across species [3–5]. The growing presence of such pollutants underscores their eventual circulation within the biosphere, where they may accumulate in soils, sediments, and food chains, leading to long-term ecological and health consequences [6, 7]. One telling example is the term "Cancer Villages," coined in China to describe areas with unusually high cancer incidence linked to polluted water bodies [8].

Water scarcity further heightens the challenge. Despite water covering 71% of Earth's surface, only 2.5% is fresh, and less than 1% is readily accessible [9]. Increasing population, rapid urbanization, climate change, and energy demands have severely stressed these freshwater reserves [10, 11]. The World Health Organization reported that as of 2019, approximately 785 million people lack access to basic drinking water services, and projections estimate that by 2025, half the global population will suffer water stress [12]. Clearly, managing wastewater effectively is essential for maintaining water quality and conserving limited freshwater resources. The United Nations, recognizing the role of clean water in human well-being and sustainable development, has prioritized water and sanitation as one of its 17 Sustainable Development Goals (SDGs) [13]. Reducing pollutant loads entering natural water bodies is crucial to meet these goals. This can be achieved by treating natural water sources before human use or, more efficiently, by treating wastewater before its release into the environment [14, 15].

Effective wastewater treatment is multifaceted. Different contaminants require different treatment strategies. Suspended solids, for example, may be removed by sedimentation, while toxic dissolved substances such as heavy metals and synthetic organics often require more advanced processes. Therefore, treatment systems typically involve a sequence of processes, tailored to address various pollutant types, to achieve regulatory compliance [11, 16]. The kinds of wastewater vary depending on the source: domestic, industrial, agricultural, and urban runoff, and contain contaminants like dyes, oils, fertilizers, and detergents [17–19]. These pollutants cause a wide range of health issues, including respiratory disorders, cancer, and heart disease. Proper pretreatment and disposal are not optional but critical to public health.

Understanding wastewater's seasonal and chemical variability is vital in designing wastewater treatment plants. Parameters such as BOD, COD, TSS, nitrogen, phosphorus, and pH help determine the appropriate treatment technologies [20, 21]. Meanwhile, the global push for sustainable water use has driven research into innovative technologies. However, traditional methods often fail to address energy efficiency, sludge volume, and removal of emerging contaminants [11, 22]. Researchers have explored advanced processes like membrane filtration, advanced oxidation, and integrated biological systems to address these gaps. These methods offer higher efficiency and broader contaminant removal potential, although their widespread implementation remains a challenge [23–27]. Moreover, emerging research emphasizes using artificial intelligence and machine learning to optimize treatment processes in real time [28].

In conclusion, wastewater treatment is not merely an environmental necessity but a public health imperative and a pillar of sustainable development. Treatment plants help preserve aquatic ecosystems, safeguard human health, and support economic resilience by preventing the discharge of harmful pollutants and recovering valuable resources like clean water. As global pressures on water resources intensify, the need for advanced, sustainable, and inclusive wastewater management systems becomes ever more critical.

1.2 Key Terminologies for Wastewater Treatment

To understand the essential aspects of wastewater treatment processes, it's important to become familiar with key technical terms frequently used. Some commonly used terms are: influent, effluent, BOD, COD, HRT, TSS, MLSS, etc. These terms are essential to define the terminology of wastewater management and reflect the parameters used to evaluate the efficiency of various processes.

(i) *Influent* refers to the raw or partially treated wastewater that enters a treatment plant. It is the starting point of the treatment process, and its characteristics (such as flow rate and pollutant load) play a crucial role in determining the design and operational needs of the treatment system.

(ii) *Effluent* is the treated water that flows out of a treatment plant after undergoing physical, chemical, or biological purification. Unlike influent, effluent has significantly lower pollutant levels and can be discharged into receiving water bodies.

(iii) *Biological Oxygen Demand (BOD)* is the amount of oxygen microorganisms consume while breaking down biodegradable organic matter in aerobic conditions. It is a key indicator of wastewater's organic strength or pollution level.

(iv) *Chemical Oxygen Demand (COD)* measures the oxygen required to chemically oxidize all organic and inorganic compounds in a water sample. COD is expressed in milligrams per liter (mg/L) and is used to estimate the total pollutant load when biological testing (like BOD) takes too long.

(v) *Total Suspended Solids (TSS)* represent the solid particles suspended in water that are not dissolved and can be captured by filtration. These include silt, organic debris, industrial waste, and sewage solids. High levels of TSS can adversely affect water quality and aquatic ecosystems.

(vi) *Volatile Solids (VS)* are the portion of solids in wastewater that can be vaporized when heated to about 550 °C. Primarily composed of organic material, volatile solids indicate biological activity and are central to sludge treatment and management processes.

(vii) *Hydraulic Retention Time (HRT)* is the average time wastewater remains in a treatment unit, such as a bioreactor or sedimentation tank. It is calculated by dividing the volume of the treatment unit by the influent flow rate and is crucial for ensuring sufficient treatment time.

(viii) *Mixed Liquor Suspended Solids (MLSS)* refers to the concentration of suspended solids in the aeration tank of a biological treatment process. It includes organic and inorganic matter, especially the microorganisms responsible for breaking down pollutants.

(ix) *Mixed Liquor Volatile Suspended Solids (MLVSS)* is a subset of MLSS, comprising the mixed liquor's volatile (mainly organic) portion. It serves as a measure of the active microbial population and is vital for assessing biological treatment efficiency.

(x) *Sludge* is the semi-solid byproduct generated during both primary and secondary treatment. It consists of concentrated organic and inorganic matter removed from the wastewater. Proper sludge handling and disposal are essential to minimize environmental risks and ensure regulatory compliance.

1.3 Sources and Characteristics of Wastewater

Wastewater can originate from various sources, including households, industries, stormwater runoff, agricultural operations, and institutional facilities. Figure 1.1 shows schematic representation of major sources of wastewater. This polluted water, which usually contains $< 1\%$ solids and the rest water, poses serious threats to public health and the environment if untreated. The sources of wastewater can be broadly categorized into five major types [29].

1. **Domestic or municipal wastewater**: This is generated from households, commercial establishments, and public facilities such as schools and hospitals. It includes greywater (from kitchens, laundry, etc.) and blackwater (from toilets), containing pathogens, detergents, and solid waste.
2. **Industrial wastewater**: It stems from sector-specific processes and may include heavy metals, acids, alkalis, detergents, and hazardous organic compounds. The nature and strength of this wastewater vary depending on the industry, such as textile dyeing, chemical manufacturing, or metal processing.
3. **Landfill leachate**: Storm-water runoff, especially in urban areas, is a significant source of pollution as it collects oils, microplastics, and debris from roads and roofs. Similarly, infiltration/inflow (I/I) occurs when groundwater enters the sewage system through pipe leaks, further diluting and complicating treatment.

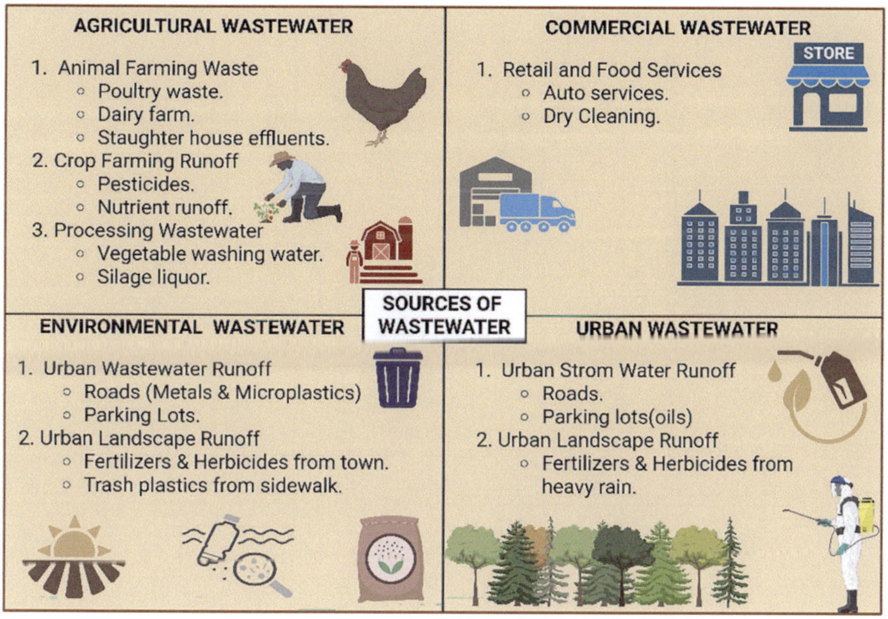

Fig. 1.1 Different sources of wastewater

4. **Agricultural wastewater**: It is another major contributor, comprising return flows from irrigation, and runoff laden with fertilizers, pesticides, and other toxic agrochemicals. This source is particularly concerning due to its diffuse nature and the high levels of nitrogen and phosphorus it introduces into water bodies.
5. **Hospital and pharmaceutical wastewater**: It contains persistent micro-pollutants and antibiotic-resistant bacteria, complicating conventional treatment methods and increasing the risk of environmental contamination.

Typical characteristics of wastewater from different industrial and domestic sources are given in Table 1.1.

1.4 Overview and Objective of Wastewater Treatment Processes

Wastewater treatment involves a series of physical, chemical, and biological processes designed to remove pollutants from the influent water progressively. These processes are typically grouped into three main stages: preliminary/primary, secondary (biological), and tertiary (advanced) treatment [38] (Fig. 1.2).

Table 1.1 Typical characteristics of wastewater from different sources

Sr. No	Source	pH	BOD (mg/L)	COD (mg/L)	TSS (mg/L)	TDS (mg/L)	Oil and grease (mg/L)	References
1	Textile	5.6–9.0	50–550	250–8000	100–700	5000–10,000	–	[30]
2	Slaughterhouse	4.9–8.1	150–4635	500–15,900	270–6400	–	–	[31]
3	Dairy industry	7.2–7.5	1300–1600	2500–3000	72,000–80,000	–	–	[32]
4	Brewery wastewater	4.6–7.3	1609–3980	1096–8926	530–3728	–	–	[33]
5	Pharmaceutical industry	7.2–8.5	480–1000	2000–3500	48–145	–	–	[34]
6	Petroleum refining	7.8–8.3	14,600–44,300	26,700–74,800	940–2470	41,600–188,000	623–9950	[35]
7	Coke oven wastewater	6.8–8.2	510–1360	930–3120	–	–	–	[36]
8	Municipal wastewater	7–8	230–560	500–1200	250–600	–		[37]

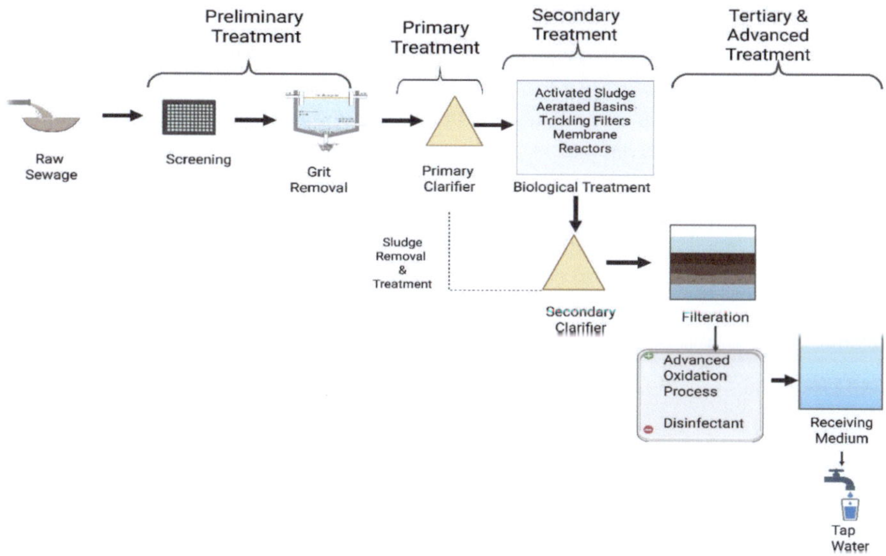

Fig. 1.2 Different stages in wastewater treatment plant

1. Preliminary treatment

Preliminary treatment or pretreatment is the initial stage and focuses on removing large solids and inert materials that could damage or hinder subsequent treatment units. The most common operations include screening, grit removal, and flow equalization. Screening involves using bar screens or fine mesh to capture floating debris such as plastics, rags, and sticks. Grit chambers then settle out heavier inorganic particles like sand and gravel that enter through storm-water or domestic waste. Flow equalization basins are used to buffer variations in flow and load, ensuring downstream units operate efficiently. This stage is critical in protecting mechanical components from clogging, abrasion, or wear.

2. Primary Treatment

Primary treatment primarily uses physical processes like sedimentation to remove suspended solids and reduce the organic load of the wastewater. In a primary settling tank (or clarifier), wastewater is held in a quiescent state to allow settling solids (like sludge) to sink to the bottom and scum or oils to float to the top for removal. This process can remove around 50–70% of TSS and 25–40% of BOD. While it doesn't significantly reduce nutrients or pathogens, primary treatment greatly reduces the burden on biological systems in the next stage.

3. Secondary Treatment

Table 1.2 Comparative analysis of wastewater treatment stages

	Primary treatment	Secondary treatment	Tertiary treatment
Objective	Remove solids that can settle and floating material	Biologically degrade dissolved organic matter and solid reduction	Remove residual suspended solids, nutrients, and pathogens
Constituents targeted	Suspended solids, large debris, floating oil, and grease	BOD, suspended solids, dissolved organics	Nitrogen, phosphorus, pathogens, and micro pollutants
% BOD Removal	25–50%	80–90%	Up to 99% (post-secondary)
% COD Removal	20–30%	40–70%	90–99%
% TSS Removal	50–70%	80–90%	> 99%
Important processes	Screening, grit removal, sedimentation, Flocculation	Activated sludge, aerated basin, trickling filters, bio-towers	Membranes, nutrient removal, and Advanced oxidation processes
Energy requirement	Low	Moderate to high	Moderate to high

Secondary treatment uses biological processes to remove dissolved and colloidal organic matter. The most widely used methods include the activated sludge process, trickling filters, and oxidation ponds. These systems rely on microbial populations to consume organic pollutants under aerobic (oxygen-rich) conditions. The activated sludge system, for instance, involves aeration tanks where microorganisms digest organic material, followed by secondary clarifiers where biomass settles and is either returned or removed as waste sludge. This stage can remove 80–90% of BOD and TSS and significantly reduce pathogens. It is a cornerstone of biological wastewater treatment.

4. Tertiary Treatment

Tertiary (or advanced) treatment is designed to polish the effluent by removing remaining contaminants not addressed in the earlier stages. This may include nutrients (nitrogen and phosphorus), residual BOD, TSS, dissolved solids, heavy metals, and micro-pollutants. Standard technologies used are sand filtration, membrane filtration, chemical coagulation, activated carbon adsorption, and disinfection (using chlorine, ozone, or UV light). In some facilities, nutrient removal is accomplished biologically through nitrification–denitrification and enhanced biological phosphorus removal. Tertiary treatment is essential for water reuse, discharge into sensitive environments, or when regulations mandate high effluent quality. Typical characteristics of different wastewater treatment stage are presented in (Table 1.2).

References

1. Industrial Wastewater and Its Toxic Effects. In *Biological Treatment of Industrial Wastewater*; The Royal Society of Chemistry, 2021; pp 1–14. https://doi.org/10.1039/9781839165399-00001.
2. Bilal, M.; Shah, J. A.; Ashfaq, T.; Gardazi, S. M. H.; Tahir, A. A.; Pervez, A.; Haroon, H.; Mahmood, Q. Waste Biomass Adsorbents for Copper Removal from Industrial Wastewater—A Review. *J. Hazard. Mater.* **2013**, *263*, 322–333. https://doi.org/10.1016/j.jhazmat.2013.07.071.
3. Environmental Contamination by Heavy Metals. In *Heavy Metals*; InTech, 2018. https://doi.org/10.5772/intechopen.76082.
4. Richardson, S. D.; Ternes, T. A. Water Analysis: Emerging Contaminants and Current Issues. *Anal. Chem.* **2018**, *90* (1), 398–428. https://doi.org/10.1021/acs.analchem.7b04577.
5. Role of Industries in Water Scarcity and Its Adverse Effects on Environment and Human Health. In *Environmental Concerns and Sustainable Development*; Springer Singapore: Singapore, 2020; pp 235–256. https://doi.org/10.1007/978-981-13-5889-0_12.
6. Sharma, N.; Singhvi, R. Effects of Chemical Fertilizers and Pesticides on Human Health and Environment: A Review. *Int. J. Agric. Environ. Biotechnol.* **2017**, *10* (6), 675. https://doi.org/10.5958/2230-732x.2017.00083.3.
7. Water Contamination by Heavy Metals and Their Toxic Effect on Aquaculture and Human Health through Food Chain. *Lett. Appl. NanoBioScience* **2020**, *10* (2), 2148–2166. https://doi.org/10.33263/lianbs102.21482166.
8. Cui, X.; Cheng, H.; Sun, H.; Huang, J.; Huang, D.; Zhang, Q. Human Health and Environment: Spatiotemporal Variation of Chinese Cancer Villages and Its Contributing Factors. *Ecol. Eng.* **2020**, *158*, 106075. https://doi.org/10.1016/j.ecoleng.2020.106075.
9. Anastasi, C. *Desalination and Water Security*, 1st ed.; Routledge: London, 2023. https://doi.org/10.4324/9781003334224.
10. Singh, V. *Textbook of Environment and Ecology*; Springer Nature Singapore: Singapore, 2024. https://doi.org/10.1007/978-981-99-8846-4.
11. Saravanan, A.; Senthil Kumar, P.; Jeevanantham, S.; Karishma, S.; Tajsabreen, B.; Yaashikaa, P. R.; Reshma, B. Effective Water/Wastewater Treatment Methodologies for Toxic Pollutants Removal: Processes and Applications towards Sustainable Development. *Chemosphere* **2021**, *280*, 130595. https://doi.org/10.1016/j.chemosphere.2021.130595.
12. Chai, W. S.; Cheun, J. Y.; Kumar, P. S.; Mubashir, M.; Majeed, Z.; Banat, F.; Ho, S.-H.; Show, P. L. A Review on Conventional and Novel Materials towards Heavy Metal Adsorption in Wastewater Treatment Application. *J. Clean. Prod.* **2021**, *296*, 126589. https://doi.org/10.1016/j.jclepro.2021.126589.
13. Photo-Oxidation Technologies for Advanced Water Treatment. In *Applied Environmental Science and Engineering for a Sustainable Future*; Springer International Publishing: Cham, 2020; pp 221–255. https://doi.org/10.1007/978-3-030-29840-1_10.
14. Chai, W. S.; Tan, W. G.; Halimatul Munawaroh, H. S.; Gupta, V. K.; Ho, S.-H.; Show, P. L. Multifaceted Roles of Microalgae in the Application of Wastewater Biotreatment: A Review. *Environ. Pollut.* **2021**, *269*, 116236. https://doi.org/10.1016/j.envpol.2020.116236.
15. Zhao, L.; Deng, J.; Sun, P.; Liu, J.; Ji, Y.; Nakada, N.; Qiao, Z.; Tanaka, H.; Yang, Y. Nanomaterials for Treating Emerging Contaminants in Water by Adsorption and Photocatalysis: Systematic Review and Bibliometric Analysis. *Sci. Total Environ.* **2018**, *627*, 1253–1263. https://doi.org/10.1016/j.scitotenv.2018.02.006.

16. Anbia, M.; Mohammadi, N.; Mohammadi, K. Fast and Efficient Mesoporous Adsorbents for the Separation of Toxic Compounds from Aqueous Media. *J. Hazard. Mater.* **2010**, *176* (1–3), 965–972. https://doi.org/10.1016/j.jhazmat.2009.11.135.
17. Ajmal, Z.; Usman, M.; Anastopoulos, I.; Qadeer, A.; Zhu, R.; Wakeel, A.; Dong, R. Use of Nano-/Micro-Magnetite for Abatement of Cadmium and Lead Contamination. *J. Environ. Manage.* **2020**, *264*, 110477. https://doi.org/10.1016/j.jenvman.2020.110477.
18. Essekri, A.; Hsini, A.; Naciri, Y.; Laabd, M.; Ajmal, Z.; El Ouardi, M.; Ait Addi, A.; Albourine, A. Novel Citric Acid-Functionalized Brown Algae with a High Removal Efficiency of Crystal Violet Dye from Colored Wastewaters: Insights into Equilibrium, Adsorption Mechanism, and Reusability. *Int. J. Phytoremediation* **2021**, *23* (4), 336–346. https://doi.org/10.1080/15226514.2020.1813686.
19. Hsini, A.; Naciri, Y.; Laabd, M.; El Ouardi, M.; Ajmal, Z.; Lakhmiri, R.; Boukherroub, R.; Albourine, A. Synthesis and Characterization of Arginine-Doped Polyaniline/Walnut Shell Hybrid Composite with Superior Clean-up Ability for Chromium (VI) from Aqueous Media: Equilibrium, Reusability and Process Optimization. *J. Mol. Liq.* **2020**, *316*, 113832. https://doi.org/10.1016/j.molliq.2020.113832.
20. Vishal, V. S. Seasonal Variations in the Performance of the Amberpet Wastewater Treatment Plant: Assessing the Impact of Temperature and Dissolved Oxygen., International Institute of Information Technology, Hyderabad, 2024.
21. Emerging Trends in Wastewater Treatment Technologies: The Current Perspective. In *Promising Techniques for Wastewater Treatment and Water Quality Assessment*; IntechOpen, 2021. https://doi.org/10.5772/intechopen.93898.
22. Agrawal, R.; Agrawal, S.; Samadhiya, A.; Kumar, A.; Luthra, S.; Jain, V. Adoption of Green Finance and Green Innovation for Achieving Circularity: An Exploratory Review and Future Directions. *Geosci. Front.* **2024**, *15* (4), 101669. https://doi.org/10.1016/j.gsf.2023.101669.
23. Hybrid Membrane Processes in Advanced Wastewater Treatment. In *Current Trends and Future Developments on (Bio-) Membranes*; Elsevier, 2024; pp 811–844. https://doi.org/10.1016/b978-0-323-90258-8.00007-9.
24. Emerging Nanoenhanced Membrane-Based Hybrid Processes for Complex Industrial Wastewater Treatment. In *Membrane-Based Hybrid Processes for Wastewater Treatment*; Elsevier, 2021; pp 633–656. https://doi.org/10.1016/b978-0-12-823804-2.00024-0.
25. Adeoye, J. B.; Tan, Y. H.; Lau, S. Y.; Tan, Y. Y.; Chiong, T.; Mubarak, N. M.; Khalid, M. Advanced Oxidation and Biological Integrated Processes for Pharmaceutical Wastewater Treatment: A Review. *J. Environ. Manage.* **2024**, *353*, 120170. https://doi.org/10.1016/j.jenvman.2024.120170.
26. Long, F.; Ghani, D.; Huang, R.; Zhao, C. Versatile Electrode Materials Applied in the Electrochemical Advanced Oxidation Processes for Wastewater Treatment: A Systematic Review. *Sep. Purif. Technol.* **2025**, *354*, 128725. https://doi.org/10.1016/j.seppur.2024.128725.
27. Babu Ponnusami, A.; Sinha, S.; Ashokan, H.; V Paul, M.; Hariharan, S. P.; Arun, J.; Gopinath, K. P.; Hoang Le, Q.; Pugazhendhi, A. Advanced Oxidation Process (AOP) Combined Biological Process for Wastewater Treatment: A Review on Advancements, Feasibility and Practicability of Combined Techniques. *Environ. Res.* **2023**, *237*, 116944. https://doi.org/10.1016/j.envres.2023.116944.
28. Wang, Y.; Cheng, Y.; Liu, H.; Guo, Q.; Dai, C.; Zhao, M.; Liu, D. A Review on Applications of Artificial Intelligence in Wastewater Treatment. *Sustainability* **2023**, *15* (18), 13557. https://doi.org/10.3390/su151813557.
29. Sources, Classifications, Constituents, and Available Treatment Technologies for Various Types of Wastewater: An Overview. In *Aquananotechnology*; Elsevier, 2021; pp 11–46. https://doi.org/10.1016/b978-0-12-821141-0.00019-7.

30. Fazal, T.; Mushtaq, A.; Rehman, F.; Ullah Khan, A.; Rashid, N.; Farooq, W.; Rehman, M. S. U.; Xu, J. Bioremediation of Textile Wastewater and Successive Biodiesel Production Using Microalgae. *Renew. Sustain. Energy Rev.* **2018**, *82*, 3107–3126. https://doi.org/10.1016/j.rser.2017.10.029.
31. Bustillo-Lecompte, C. F.; Mehrvar, M. Slaughterhouse Wastewater Characteristics, Treatment, and Management in the Meat Processing Industry: A Review on Trends and Advances. *J. Environ. Manage.* **2015**, *161*, 287–302. https://doi.org/10.1016/j.jenvman.2015.07.008.
32. Demirel, B.; Yenigun, O.; Onay, T. T. Anaerobic Treatment of Dairy Wastewaters: A Review. *Process Biochem.* **2005**, *40* (8), 2583–2595. https://doi.org/10.1016/j.procbio.2004.12.015.
33. Durban University of Technology, Durban, South Africa; Enitan, A. M.; Adeyemo, J.; Kumari, S. K.; Swalaha, F. M.; Bux, F. Characterization of Brewery Wastewater Composition. *Int. J. Environ. Ecol. Eng.* **2015**, *9* (9), 1073–1076. https://doi.org/10.51415/10321/2992.
34. Wei, X.; Li, B.; Zhao, S.; Wang, L.; Zhang, H.; Li, C.; Wang, S. Mixed Pharmaceutical Wastewater Treatment by Integrated Membrane-Aerated Biofilm Reactor (MABR) System – A Pilot-Scale Study. *Bioresour. Technol.* **2012**, *122*, 189–195. https://doi.org/10.1016/j.biortech.2012.06.041.
35. Eldos, H. I.; Khan, M.; Zouari, N.; Saeed, S.; Al-Ghouti, M. A. Characterization and Assessment of Process Water from Oil and Gas Production: A Case Study of Process Wastewater in Qatar. *Case Stud. Chem. Environ. Eng.* **2022**, *6*, 100210. https://doi.org/10.1016/j.cscee.2022.100210.
36. Chang, E.-E.; Hsing, H.-J.; Chiang, P.-C.; Chen, M.-Y.; Shyng, J.-Y. The Chemical and Biological Characteristics of Coke-Oven Wastewater by Ozonation. *J. Hazard. Mater.* **2008**, *156* (1–3), 560–567. https://doi.org/10.1016/j.jhazmat.2007.12.106.
37. Wastewater Characteristics. In *Biological Wastewater Treatment: Principles, Modeling and Design*; Volcke, E. I. P., Solon, K., Comeau, Y., Henze, M., Eds.; IWA Publishing, 2020; pp 77–110. https://doi.org/10.2166/9781789060362_0077.
38. Riffat, R.; Husnain, T. *Fundamentals of Wastewater Treatment and Engineering*, Second edition.; CRC Press: Abingdon, Oxon ; Boca Raton, FL, 2022.

Preliminary and Primary Treatment Processes 2

Abstract

This Chapter addresses the crucial processes of preliminary and primary wastewater treatment, focusing on the removal of coarse solids and preparation for advanced treatment stages. It begins with an overview of initial therapy, highlighting operations like screening and filtering, and detailing components such as bar screens, fine screens, grit chambers, and comminutors. Flow equalization is also key to maintaining consistent hydraulic and organic loading. The chapter then covers coagulation and flocculation, explaining the mechanisms, chemical agents used, and design considerations for optimizing performance. It introduces Chemically Enhanced Primary Treatment to improve pollutant removal efficiency. The latter sections explore primary sedimentation, including settling phenomena, sedimentation theory, and tank design. It also touches on advanced topics like equations for settling velocity and using Computational Fluid Dynamics in design. Overall, this chapter provides essential knowledge and practical insights into the initial stages of wastewater treatment, setting the stage for secondary and tertiary strategies.

2.1 Overview of Preliminary Treatment

Preliminary treatment is the first stage in the wastewater treatment train and is a protective barrier for downstream processes. Its primary function is to remove large and coarse materials, such as rags, plastics, grit, and grease, which can cause damage to equipment, clog pipelines, or reduce the efficiency of later stages. This section introduces the key components of preliminary treatment, including screening units, grit chambers, and flow

A. H. Bari et al., *Wastewater Treatment*, Synthesis Lectures on Sustainable Development,
https://doi.org/10.1007/978-3-032-07585-7_2

equalization tanks. By eliminating heavy debris and regulating flow fluctuations, preliminary treatment ensures smoother operation and prolongs the lifespan of biological and mechanical systems further along the treatment path.

2.2 Screening and Filtering

Screening and filtering are the first steps in the physical pre-treatment of wastewater. These operations aim to remove large solids such as rags, sticks, plastics, and other debris that could damage downstream equipment or obstruct flow. Screens are essential in preliminary wastewater treatment to remove large and fine solids. The most common screen types include trash racks, bar screens, fine screens, and micro screens [1]. Trash racks are the coarsest type of screen, typically installed at the intake of treatment plants or pumping stations. They have large openings (typically > 50 mm) and are intended to intercept bulky objects such as branches, plastics, or rags. Bar screens are the next screening stage and consist of vertical or inclined bars with spacing usually ranging from 6 to 150 mm. They effectively remove larger solids and are often divided into coarse and fine bar screens. Bar screens can be manually or mechanically cleaned. Fine screens are designed to remove finer particulate matter that passes through bar screens. Constructed using perforated plates, wire mesh, or wedge wire, they typically have openings between 0.5 and 6 mm. These screens effectively remove fine debris such as seeds, hair, and food particles. Micro screens (or microstrainers) are precision screening devices, often in the form of rotating drum filters, with tiny openings (<0.5 mm). They are used in advanced or tertiary treatment stages to remove algae or wonderful suspended solids.

2.2.1 Bar Screen

Coarse screens are smaller-sized opening screens used at the very beginning of a wastewater treatment plant to remove large floating and suspended solids, such as rags, plastic, and wood. This screen consists of a series of vertical or slightly inclined bars fixed at regular intervals across a channel through which the wastewater flows. Large debris is trapped on the bars as the water passes through the screen. These screens can be manually cleaned or mechanically cleaned [1]. Manually cleaned bar screens may be used in small-sized wastewater treatment plants. In a mechanically cleaned bar screen, a mechanical rake or comb system is fitted to ensure continuous operation and prevent clogging, which moves along the screen at regular intervals or in response to flow conditions. This rake automatically lifts and removes the collected debris, which is then discharged into a collection bin or conveyor for further disposal. Mechanically cleaned bar screens reduce labor, prevent overflow, and protect downstream processes from damage caused by large solids. Effective screening reduces downstream clogging, protects pumps and mixers, and reduces organic load entering the treatment plant.

2.2.2 Fine Screen

Fine screens are integral to the preliminary or intermediate stages of wastewater treatment, designed to remove smaller solids that pass through coarse screens. These screens typically have openings less than 6 mm, with some as fine as 0.5 mm [1], and are essential in protecting downstream equipment and enhancing the efficiency of biological processes. Fine screens are used where wastewater contains minimal coarse solids or requires recovering fine, valuable materials from industrial streams. Unlike bar racks, fine screens often use mesh cloths made from stainless steel or other corrosion-resistant materials. These may be mounted on rotating drums or discs that transport retained solids out of the flow for removal by sprays or brushes. Disc-type screens use vertically mounted hoops rotating through the water, while drum-type screens are horizontally oriented cylinders through which wastewater flows, collecting solids either inside or outside the screen cloth.

Manufacturing techniques for fine screens vary, including slotted perforated plates (0.8–2.4 mm), wedge-shaped bars, looped wire, and woven wire cloth with specific openings tailored to treatment needs. Several commercial designs are available [1]:

(i) Static wedgewire screens, with 0.2–1.2 mm openings and high throughput (400–1200 $L/m^2 \cdot min$)
(ii) Stair screens, with moving and fixed steps that lift solids mechanically (3–6 mm openings),
(iii) Rotary drum screens can be installed in various flow directions and are available in several sizes.

A rotary drum screen, or a revolving drum screen (Fig. 2.1a), removes fine suspended solids from wastewater before they enter sensitive treatment units like membrane bioreactors. This equipment consists of a cylindrical drum covered with a perforated or wedge-wire screen and mounted to rotate slowly on its axis. Wastewater is introduced into the drum, either on the inside or the outside, depending on the design. Solid particles are trapped on the screen surface as the drum rotates while the clarified water passes through. The trapped solids are then dislodged from the screen surface by internal spray jets, brushes, or scrapers and are discharged into a collection trough. The continuous rotation and cleaning ensure high screening efficiency and reduce the risk of clogging, especially when dealing with wastewater with a high content of fine debris.

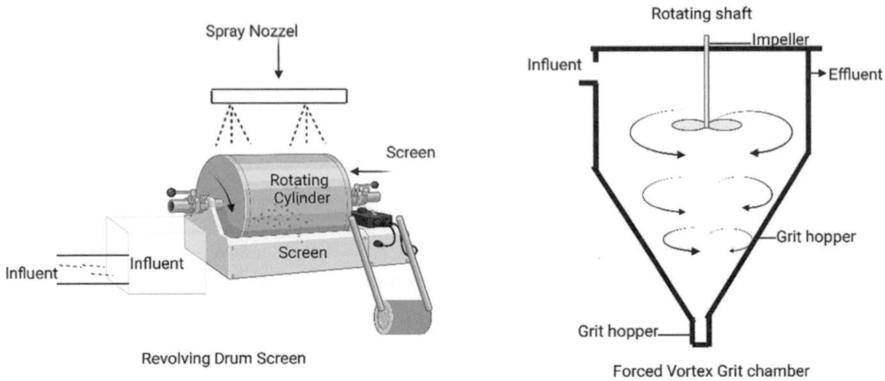

Fig. 2.1 Schematic of different equipment's used for screening. **a** Revolving drum screen, **b** Forced vortex grit chamber

2.2.3 Grit Chamber

Grit chambers are units used in preliminary wastewater treatment to remove heavy inorganic solids such as sand, gravel, coffee grounds, eggshells, and other non-biodegradable or slowly biodegradable materials. These chambers are positioned after the screening stage and before primary sedimentation. They mainly aim to protect downstream equipment from abrasion, prevent pipe blockages, and reduce grit accumulation in digesters. Grit particles have a nominal diameter of 0.15–0.20 mm and a specific gravity of around 2.65. Settling velocities for grit typically range from 1.3 to 2.3 cm/s, depending on size and type [1].

There are three main types of grit chambers:

Horizontal Flow Grit Chambers—These are rectangular open channels designed to maintain constant flow velocity for grit settling. They are often found in older plants and may require manual or mechanical cleaning.

Aerated Grit Chambers—Common in modern facilities, these chambers use air to create a spiral flow, keeping lighter organics suspended while heavier grit settles. Advantages include reduced odor and minimal headloss, but they have higher energy demands and maintenance needs.

Forced Vortex Grit Chambers—These systems create a circular flow that lifts organic matter while grit settles into a hopper. They are compact and efficient under high flows, but may have grit compaction challenges. Figure 2.1b shows a schematic of a forced vortex grit chamber. In this chamber, wastewater enters tangentially, which induces a vortex motion in the tank along with a mechanical agitator or turbine. This artificial circulation creates a centrifugal force that pushes heavier grit particles outward and downward toward the bottom of the chamber, where they settle into a grit hopper. Meanwhile, lighter organic particles remain suspended in the flow and exit the tank with the effluent. The grit is continuously or periodically removed using scraping mechanisms, airlifts, or pumps. This method is highly effective for grit separation and minimizes organic contamination in the removed material, making it ideal for modern municipal treatment plants.

2.2.4 Comminutor

Comminutors are mechanical devices used in preliminary wastewater treatment to grind or shred solid materials present in the wastewater without actually removing them from the flow. These units are typically installed in channels where wastewater passes through rotating cutting elements. As large solids enter the comminutor, they are captured and reduced in size by shearing action between rotating and stationary blades or grinding teeth. The shredded material remains in the wastewater, now in a form that is less likely to clog pipes, pumps, or downstream treatment units. Comminutors are especially useful when bar screens are not preferred or there is limited space, as they combine screening and shredding functions in a compact and low-maintenance system. Comminutors are available in a range of sizes. Smaller units (as small as 10.16 cm or 4 in. diameter) serve low-flow applications, while larger units (up to 137.16 cm or 54 in. diameter) can process flows up to 3.15 m^3/s (72 MGD). They typically use slot widths of 0.635–0.952 cm and require power ranging from 186 W (¼ hp) to 1491 W (2 hp) [2].

2.3 Flow Equalization

Equalization is a process in which variations in wastewater flow and composition are balanced to provide more consistent conditions for downstream treatment units. Flow equalization tanks are essential components in wastewater treatment systems, particularly in addressing the challenges caused by fluctuating influent flow rates and pollutant concentrations. Variations in daily influent flows significantly impact the design and operation of wastewater treatment plants. Such variations can lead to increased capital costs due to the need for oversizing unit processes to accommodate peak flows [3]. Also, the fluctuating hydraulic loads can disrupt the performance of secondary settling tanks (SSTs), causing sludge overflow and poor effluent quality [4]. These variations also complicate oxygenation, reducing the efficiency of nitrification and COD removal. Under- or over-aeration can result in energy waste, deteriorated sludge characteristics, and the growth of filamentous organisms that hinder settling. Flow equalization helps to buffer these variations, stabilizing influent conditions and thus reducing operational difficulties.

While flow equalization tanks provide clear benefits, the extent of their effectiveness depends heavily on the process parameters monitored and operational context. The traditional indicators like COD or BOD may not accurately reflect improvements from equalization due to their insensitivity under dynamic conditions [3]. Benefits become more evident in systems with long sludge ages or when SSTs are cyclically overloaded. When well-designed and properly operated, equalization tanks can significantly enhance plant performance. They reduce hydraulic disturbances, improve SST and Primary Settling Tank efficiency, allow for extended sludge ages for better nitrification, and simplify operations such as aeration and chemical dosing [3, 4]. Flow equalization also reduces the variability of solids entering reactors, thus minimizing denitrification challenges. Despite a marginal increase in power consumption (2–6%), the overall effect remains positive [5]. Therefore, when implemented under suitable conditions with proper monitoring and operational strategies, flow equalization tanks are valuable in ensuring stable and efficient Wastewater treatment plant operation.

There are two primary types of flow equalization: in-line and side-line or off-line (Fig. 2.2). In in-line equalization, all wastewater flows directly through the equalization tank before entering subsequent treatment stages, ensuring uniformity in flow and load. In side-line equalization, only a portion of the peak flow is diverted into the tank during high-flow periods and released gradually during low-flow conditions to stabilize the overall loading on the treatment plant. The design of equalization tanks involves several considerations. The required volume is typically based on 6–12 h of average flow, though it may vary depending on flow variability. Aeration is often included to prevent septicity and maintain solids in suspension, while proper mixing is essential to avoid solids settling and stratification. Automated control systems, comprising level sensors, flow meters, and programmable logic controllers (PLCs), may regulate inflow and outflow effectively.

Key design parameters include the required retention time, which is determined by flow variability and the level of damping needed, and the shape of the tank, which may be

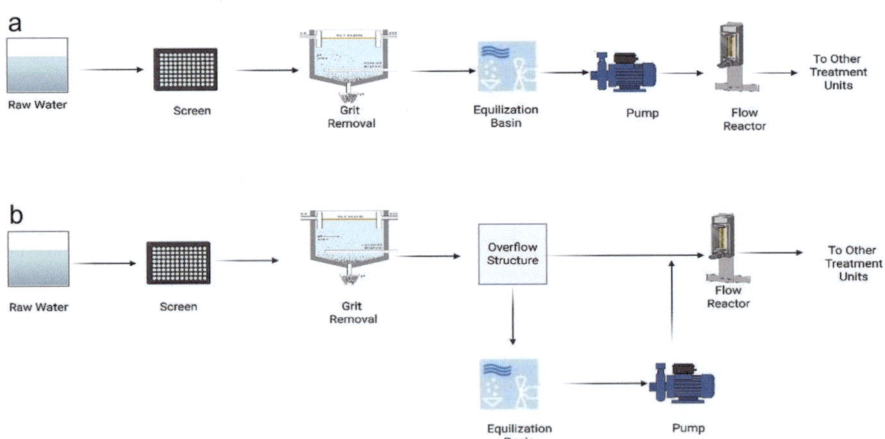

Fig. 2.2 Types of flow equalization **a** In-line **b** side-line

rectangular or circular, depending on site constraints and space availability. When appropriately designed, equalization tanks enhance operational efficiency, reduce maintenance costs, and significantly improve the robustness and reliability of biological treatment processes.

2.4 Coagulation and Flocculation Processes

Coagulation and flocculation are critical physico-chemical processes employed in water and wastewater treatment to remove suspended solids, colloids, and organic and inorganic contaminants. In these processes, fine particles are aggregated into larger flocs, which can settle easily. These flocs are removed through sedimentation and filtration processes, significantly improving water clarity and quality. Traditionally dependent on chemical additives, the process increasingly incorporates natural and composite materials to address environmental and health concerns. These methods serve as pre-treatment steps that significantly impact the performance of subsequent processes such as biological treatment, filtration, and disinfection.

2.4.1 Mechanism of Coagulation and Flocculation

The principles of coagulation have been understood for over a century, with early applications involving aluminum sulfate and ferric chloride in full-scale water treatment systems. Classical theories, including the Schultz–Hardy rule and Smoluchowski's collision function, emphasized overcoming the natural electrostatic repulsion between particles to enable their aggregation. In natural waters, particles typically carry a negative surface charge within the usual pH range of 5–9 [6]. This is often because the pH is above the isoelectric point (IEP) of most particles, particularly those of biological origin. Even mineral particles with a relatively high IEP can acquire a negative charge through the adsorption of natural organic matter. As a result, these particles remain colloidally stable and resist aggregation unless destabilized by external agents.

Coagulation serves to destabilize these particles, and this can occur through ways. As per classical ideas, two primary pathways for destabilization have been recognized: (1) increasing the ionic strength of the solution, which compresses the electrical double layer and reduces zeta potential, and (2) specific adsorption of counterions that neutralize the particle's surface charge. Effective additives under these conditions are salts containing highly charged cations. Another theory emphasizes that it is not the metal ions themselves, but rather their hydrolysis products, primarily responsible for destabilizing particles and initiating coagulation [7]. The coagulation follows any of the four types of coagulation mechanisms, viz., (i) double layer compression, (ii) polymer bridging, (iii) charge neutralization, and (iv) sweep coagulation. Each Mechanism plays a role depending on factors

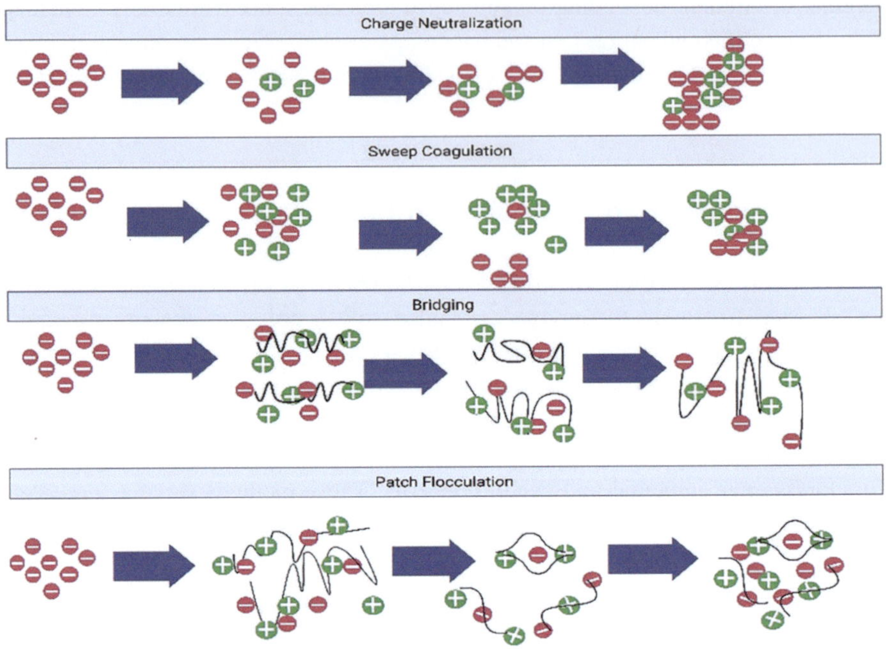

Fig. 2.3 Mechanism of coagulation

like water chemistry, coagulant type, and dosage, and in many cases, more than one Mechanism may act simultaneously (Fig. 2.3).

Once destabilized, particles can collide and aggregate to form larger flocs. This subsequent step, known as flocculation, involves gently mixing water to promote collisions among destabilized particles. These collisions form larger aggregates through mechanisms such as inter-particle bridging, physical entrapment, and van der Waals attraction. Flocculation kinetics are traditionally described by Smoluchowski's model, which Camp and Stein later refined through the introduction of the velocity gradient (G), a key parameter for understanding mixing intensity in flocculation systems [8]. According to DLVO (Derjaguin–Landau–Verwey–Overbeek) theory, the balance between attractive van der Waals forces and repulsive electrostatic interactions governs the stability of colloidal suspensions. In practice, however, non-DLVO forces, including hydrophobic interactions, steric hindrance, and hydration pressure, also play significant roles in real water systems [9]. Understanding these micro-scale interactions is increasingly important as modern research focuses on optimizing floc structure and behavior under dynamic flow conditions. The effectiveness of coagulation and flocculation thus depends not only on chemical selection and dosage but also on careful control of hydrodynamic conditions within treatment systems.

2.4.2 Types of Coagulants and Flocculants

2.4.2.1 Chemical Coagulants

Chemical coagulants are the most widely used agents in conventional water and wastewater treatment due to their high efficiency and rapid action in destabilizing colloidal particles. Chemical coagulants such as aluminum sulfate and ferric chloride have been widely used. As water quality standards became more stringent, researchers sought to enhance coagulation efficiency by developing new coagulants. These coagulants work primarily by neutralizing the negative surface charges of suspended particles, thereby reducing repulsion and promoting aggregation. Upon addition to water, they undergo hydrolysis to form metal hydroxide precipitates, which enmesh particles through a mechanism known as sweep flocculation. Introducing pre-polymerized coagulants like polyaluminum chloride (PACl) and polyferric sulfate (PFS) has enhanced floc strength, reduced dosage requirements, and improved turbidity removal [10]. Mixed-metal coagulants such as poly-aluminum-iron-sulfate were also developed, demonstrating synergistic effects in enhancing floc formation and settling characteristics. Despite their effectiveness, these chemicals can leave residual metals in treated water and produce large volumes of sludge requiring specialized disposal. Nevertheless, chemical coagulants remain the backbone of large-scale treatment due to their reliability, cost-effectiveness, and ease of application.

2.4.2.2 Composite and Magnetic Coagulants

Composite and magnetic coagulants represent a significant advancement in coagulation technology, aiming to enhance treatment efficiency while addressing the limitations of conventional chemical coagulants. Composite coagulants combine metal salts with other functional materials, such as polymers, clays, or natural extracts- to improve floc formation, strength, and pollutant removal across a broader range of water qualities. For example, polyferric sulfate combined with organic or mineral additives has superior performance in removing turbidity and organic matter. Magnetic coagulants, a subset of composite materials, incorporate magnetic nanoparticles (e.g., Fe_3O_4) into the coagulant matrix. These materials facilitate rapid settling and post-treatment sludge recovery using external magnetic fields, reducing sludge handling complexity and allowing potential reuse [11]. Modified montmorillonite clays loaded with Al_{13} species have proven effective for targeted contaminant removal, such as arsenic [12]. These advanced materials also support circular economy principles by enabling resource recovery, such as the regeneration of coagulants or the recovery of valuable metals. While promising, the high production cost and need for large-scale validation remain barriers to widespread adoption. Nonetheless, composite and magnetic coagulants offer a versatile, eco-friendly solution for modern water treatment challenges.

2.4.2.3 Natural Coagulants

Traditional chemical coagulants have drawbacks, including residual toxicity, sludge generation, and health risks, prompting a shift towards natural alternatives. Natural coagulants, particularly those derived from plants, are gaining popularity as environmentally friendly alternatives to traditional chemical coagulants. These coagulants perform by mechanisms such as polymer bridging and charge neutralization, effectively destabilizing colloidal particles [13]. Polymer bridging involves long-chain biopolymers attaching to multiple particles, thereby linking them into larger flocs, while charge neutralization uses positively charged natural polyelectrolytes to counter the negative charge of colloids. Plant-based sources—like Moringa oleifera, jackfruit seed starch, banana pith, neem leaf, and orange peel—have demonstrated high removal efficiencies for turbidity, suspended solids, and microbial contaminants [14, 15]. For example, orange peel powder and neem leaf achieved 96 and 98% turbidity removal from dairy wastewater [16], and Moringa oleifera removed more than 80% from model wastewater [17]. These coagulants are effective, safer for health, and produce less toxic sludge than alum-based treatments. However, challenges remain, including introducing organic matter that may increase microbial activity and the need for standardization and scalability for widespread use.

2.4.3 Design Considerations

The choice of coagulant plays a critical role in treatment efficiency. Factors influencing the selection of a coagulant include the type of impurities, cost, availability, sludge generation, and post-treatment requirements. Optimum dosing is essential to avoid under- and over-dosing, which can lead to poor floc formation or charge reversal. Jar tests are a conventional method for determining optimal coagulant dose, where water samples are treated with varying amounts of coagulant to observe floc formation and turbidity reduction. Designing effective coagulation and flocculation systems requires careful consideration of hydraulic and mechanical parameters. Key among these is the mixing intensity (quantified as the velocity gradient or G-value), detention time, reactor geometry, and flow conditions.

Coagulation requires rapid mixing to ensure uniform coagulant distribution and initiate particle destabilization. Rapid mix units are usually high-speed impellers designed to provide short retention times (typically 30–60 s) with high G-values (700–1000 s^{-1}). The mixing must be intense enough to disperse the coagulant evenly without breaking already-formed flocs. Flocculation, on the other hand, demands gentle mixing to promote particle collision and aggregation. Flocculators are designed with lower G-values (20–80 s^{-1}) and longer retention times (15–45 min), providing the right balance between floc growth and shear sensitivity [18].

There are three types of primary flocculators: Hydraulic Flocculators, mechanical flocculators, and Helically Coiled Tube Flocculators. Hydraulic flocculators utilize the energy from flowing water to create controlled turbulence without mechanical input. Common designs include horizontal and vertical baffled tanks, cyclone, porous plate, and impeller-type configurations. These systems are energy-efficient and low-maintenance, making them ideal for constant flow applications. Mechanical flocculators**,** in contrast, use impellers or paddles to stir water, offering precise control over mixing intensity and floc size. Though they consume more energy, they allow better control over floc size and performance under fluctuating conditions. An important design parameter is the product of G and detention time (G_t), representing the total mixing energy imparted. For optimal floc formation, Gt values typically range between 10^4 and 10^5 s, depending on water quality and treatment goal [18]. Various experimental setups and computational studies using Population balance modelling [19] and Computational Fluid Dynamics (CFD) [20] have shown that impeller geometry, tank size, and speed significantly influence floc properties and system performance. While mechanical flocculators provide high adaptability and efficiency, they are more complex, costlier, and require regular maintenance. Helically Coiled Tube Flocculators (HCTFs) are compact units that create spiral flow within coiled tubes, improving mixing while reducing shear stress on flocs. Their compact design makes them suitable for space-limited or decentralized water treatment systems. They show high turbidity removal efficiencies (up to 86.2%) and lower energy consumption, though their design and operation require precision [21, 22]. HCTFs represent a promising solution for sustainable and efficient water treatment. Thus each type of flocculator provides specific advantages tailored to different operational needs, and the choice of flocculator depends on energy use, space availability, flow variability, cost, and maintenance.

2.5 Chemically Enhanced Primary Treatment

While coagulation and flocculation are commonly used as pre-treatment strategies to aggregate colloidal particles and improve water quality, Chemically Enhanced Primary Treatment (CEPT) refers to integrating these processes directly within the primary sedimentation stage of wastewater treatment. Researchers have referred to the CEPT process using various terms, including chemically assisted primary sedimentation (CAPS), chemically enhanced primary sedimentation (CEPS), chemically enhanced mechanical treatment (CEMT), and advanced primary treatment (APT). CEPT enhances the removal of solids and pollutants by applying coagulation and flocculation principles within the primary clarifier. Unlike traditional coagulation and flocculation, which primarily aim to reduce turbidity and aid downstream filtration, CEPT focuses on carbon capture, phosphorus removal, and preventing the overloading of biological treatment units [23]. Its main objective is to redirect nutrients and pollutants, particularly organic carbon, into the produced sludge, which can be considered a valuable resource for energy recovery. In this way, CEPT serves as a pollutant removal method and a strategic tool for carbon redirection, supporting treatment approaches emphasizing energy efficiency and nutrient management

[24]. Thus, CEPT differs from conventional pre-treatment coagulation by combining chemical dosing with physical sedimentation to maximize the early-stage removal of suspended solids, organic matter, and nutrients.

CEPT enhances performance by integrating coagulation–flocculation mechanisms, such as charge neutralization, sweep flocculation, and inter-particle bridging with hydraulic design and sedimentation. Using CEPT, 80–90% of TSS, including some colloidal particles, and 50–80% of the biochemical BOD can be removed. Compared, well-designed and efficiently operated primary sedimentation tanks without chemical addition typically achieve 50–70% TSS removal and 25–40% BOD reduction [25]. Following chemical addition, rapid and slow mixing promote the formation of large, dense flocs removed in the primary clarifier. Thus, CEPT simultaneously serves as both a chemical and physical treatment. Various coagulants are central to CEPT's improved efficiency, most commonly ferric chloride, alum, and polyaluminum chloride (PACl). These chemicals precipitate organic and inorganic phosphorus, bind colloidal organics, and form dense aggregates. Anionic or cationic polymers are often added as flocculants to strengthen floc structure and speed up settling [24]. Some coagulants and polymers used in CEPT, such as PACl or long-chain organic polymers, are relatively less common in standard coagulation/flocculation applications. Pilot studies report that dosing as low as 150 mg/L of ferric chloride can yield up to 88% COD removal, > 99% phosphorus removal, and substantial reductions in TSS [26]. CEPT thus significantly reduces organic loads reaching downstream biological units, making secondary treatment more stable and efficient.

Besides high removal rates, CEPT supports improved carbon management. By capturing more biodegradable organics in primary sludge, CEPT redirects carbon toward anaerobic digestion, promoting biogas production and reducing the aeration demand during biological treatment. In terms of comparative performance, CEPT has demonstrated COD removal efficiencies of 43–95%, TSS reductions up to 99.5%, and phosphate removal exceeding 99%, depending on coagulant type and dosage [24]. These high removal rates often allow clarifier size reductions of up to 50%, offering an effective solution for plants facing space or capacity constraints [27].

However, CEPT has some limits: removal of dissolved pollutants such as hydrophilic organic micropollutants or viruses is often under 50%, necessitating tertiary polishing like filtration, ozonation, or membrane treatment when required [26]. Moreover, chemical dosing must be tightly controlled, especially under fluctuating pH, temperature, or influent conditions—to avoid over- or under-dosing. For example, PACl and Ammonium sulphate negatively impacted methane yield in anaerobic sludge digestion, and reduction was about 14.4–31.7% [28]. Additional challenges include increased primary sludge volumes, about 1.5–2 times conventional primary treatment, though it can be used for obtaining adsorbents [29]. Moreover, a key limitation of existing CEPT application studies is the lack of evaluation regarding the environmental impacts of the coagulants and flocculants used. Emerging research focuses on addressing the shortcomings of CEPT by using hybrid CEPT systems [30], magnetic coagulants to aid pollutant recovery [31], bio-based coagulants to lower chemical costs [32], and integration with membrane [33].

Nevertheless, CEPT represents a transformative enhancement of primary sedimentation: by fusing coagulation/flocculation chemistry with sedimentation engineering, it markedly improves removal of solids, organics, and nutrients, supports energy recovery, reduces plant footprint and energy use, and prepares wastewater plants for future sustainability challenges, provided its operational demands and limitations are carefully managed.

2.6 Primary Sedimentation Process

The primary sedimentation process removes suspended solids, allowing them to settle under gravity in large basins. The basin in which the flow is retarded and sedimentation takes place is called a settling tank. Suspended solids in water with specific gravity greater than water's tend to settle down by gravity. During this stage, a significant portion of the suspended solids and some organic matter are separated from the wastewater, forming a sludge layer at the bottom of the tank. The detention period is when the water is held in the settling tank.

2.6.1 Types of Settling

Theoretically, based on the mechanisms, gravitational settling can be classified into four types: discrete, flocculent, hindered (zone), and compression settling. Discrete and flocculant settlings are common in the primary sedimentation tank. However, each type of settling plays a critical role in different treatment units and reflects unique particle interactions.

Type I settling, also known as discrete particle settling, occurs in dilute suspensions containing non-flocculent particles. In this mode, individual particles maintain their identity and settle independently without interacting with one another. The process adheres to Stokes' Law, which predicts the terminal settling velocity based on particle size, density, and fluid viscosity. This type is commonly used in grit chambers to remove heavier, irregular particles such as sand and silt from raw wastewater.

Type II settling, or flocculent settling, involves dilute but floc-forming suspensions. Individual particles tend to aggregate, flocculate, or coalesce. As this increases their size and mass, their settling velocity rises, leading to a non-uniform but accelerated settling. This process is standard in primary sedimentation tanks with the aid of chemical coagulants; however, sometimes it is also carried out in the upper part of secondary sedimentary tanks, aiding in removing TSS and chemically bound organic matter.

Type III settling, or hindered or zone settling, occurs in moderate to high solid concentrations. When there is a high concentration of solids, particles are densely packed and settle as a cohesive blanket. Due to the limited relative motion among particles, the entire sludge blanket settles as a single unit. A clear interface between the solid and liquid phases can be observed, which moves downward, and this interface settling velocity is used in settling tank design. This type of settling is typical in secondary clarifiers and sludge gravity thickeners.

Finally, Type IV settling, or compression settling, happens under very high solids concentration conditions. The compression occurs due to the weight of the particles as the sludge has reached a point of compaction. The particles are in continuous contact, and further settling is only achieved through expulsing water from the compressed mass. This behavior is observed in sludge thickening units or digesters, where biological or chemical sludge is concentrated for subsequent processing or disposal.

Understanding these distinct settling mechanisms is essential for properly designing and operating sedimentation units, as each type demands specific hydraulic and structural considerations to optimize solid–liquid separation (Fig. 2.4).

2.6.2 Sedimentation Theory

Whenever a particle moves through a fluid, it is acted upon by gravitational, buoyancy, and drag forces. These forces can be expressed as:

$$F_g = mg \tag{2.1}$$

$$F_B = \rho g V_p \tag{2.2}$$

$$F_D = \frac{C_D \rho v^2 A_p}{2} \tag{2.3}$$

While gravitational force will act in the downward direction, buoyancy and drag forces act in the upward direction. The resultant force will impart acceleration to the particle. Thus, the equation of velocity is given as:

$$m\frac{dv}{dt} = F_g - F_B - F_D \tag{2.4}$$

The particle attains a constant settling velocity when the net force becomes zero. This settling velocity is the most important parameter in designing a sedimentation tank. After solving Eqs. 2.1–2.4 for net zero force,

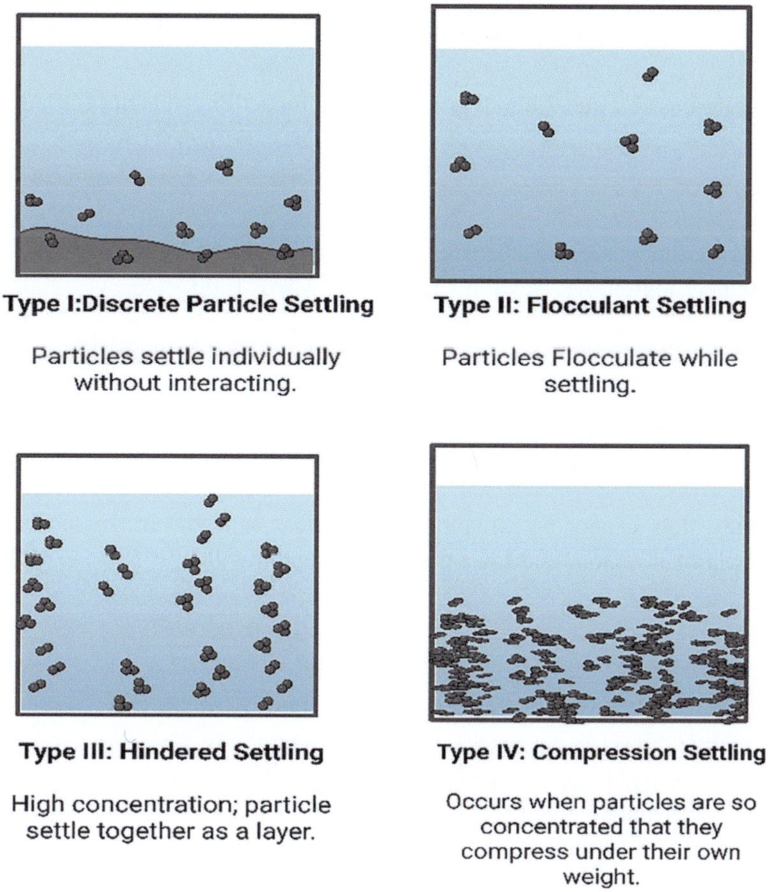

Fig. 2.4 Different types of settling

$$v_t = \sqrt{\frac{4D_p g(\rho_p - \rho)}{3C_D \rho}} \tag{2.5}$$

The value of C_D Typically depends on particle Raynolds number $N_{Re,p}$

$$N_{Re,p} = \frac{D_p \rho v_t}{\mu} \tag{2.6}$$

For the Stokes law range, i.e., the laminar region where particle Raynold's number is less than 1.0 [34]:

$$C_D = \frac{24}{N_{Re,p}} \tag{2.7}$$

$$N_{Re,p} = \frac{D_p \rho v_t}{\mu} \tag{2.8}$$

Substituting Eqs. 2.7 and 2.8 in 5 gives equation for settling velocity for laminar region

$$v_t = \frac{g D_p{}^2 (\rho_p - \rho)}{18\mu} \tag{2.9}$$

Generally, sedimentation in a settling tank typically occurs in the laminar flow region. For the turbulent region, i.e. Newton's law range, where $1000 < N_{Re,p} < 200{,}000$

$$C_D = 0.44$$

$$v_t = 1.75\sqrt{\frac{g D_p (\rho_p - \rho)}{\rho}} \tag{2.10}$$

Sometimes, flow cannot be categorized as laminar or turbulent. In such a case, it is called a transitional region, and for a transitional region with $1 < N_{Re,p} < 1{,}000$,

$$C_D = \frac{24}{N_{Re,p}} + \frac{3}{\sqrt{N_{Re,p}}} + 0.34 \tag{2.11}$$

The key objective of a sedimentation tank is to allow particles with a specific terminal settling velocity v_t to settle out before reaching the outlet. For a rectangular sedimentation tank with Length, width, and depth equal to L, B, and H, respectively, and a water flow rate of Q, the horizontal velocity v_h is:

$$v_h = \frac{Q}{BH} \tag{2.12}$$

Each discrete particle will be moving with a horizontal velocity v_h and a downward settling velocity v_t. The vector sum of these velocities gives the resultant path. From simple geometrical considerations for a rectangular tank, it can be easily shown that all particles having a settling velocity equal or greater than $\frac{Q}{BL}$ will settle down and be removed. This term is also called as surface overflow rate (SOR), or surface loading rate v_o

$$v_o = \frac{Q}{BL} = \frac{Q}{A_s} \tag{2.13}$$

The removal efficiency of a sedimentation tank for a specific particle size can be approximated as:

$$\eta = \frac{v_t}{v_o} \times 100 \tag{2.14}$$

This illustrates that the critical factor for design is the surface overflow rate v_o and not the depth H of the basin, which instead affects the volume and sludge holding capacity.

2.6.3 Sedimentation Tanks

Sedimentation tanks provide the necessary detention time and quiescent conditions for effective particle removal. Among various designs, rectangular and circular tanks are the most commonly used due to their efficiency and adaptability to different treatment plant configurations. Rectangular sedimentation tanks, typically ranging from 15 to 90 m in length, 3 to 24 m in width, and 3 to 4.9 m in depth [25], are horizontal-flow units where water enters from one end and flows linearly toward the outlet, allowing solids to settle as flow progresses. These tanks are simple in construction, provide good flow control, and are easy to build and operate, especially in large-scale treatment plants. However, they require more land area and may experience short-circuiting and uneven sludge distribution.

Circular tanks, on the other hand, operate on a radial-flow principle where water enters at the center and moves outward toward a peripheral outlet. These tanks typically range in diameter from 3 to over 60 m, with depths also between 3 and 4.9 m [25]. They are compact, require less space, and are easier to equip with mechanical sludge removal systems due to their geometry. The main drawback of circular tanks lies in their higher construction cost and potential for inefficient performance at very large diameters if flow distribution is not well managed. In smaller treatment systems, hopper-bottom tanks are sometimes used for their ability to concentrate and discharge sludge efficiently without mechanical scrapers, especially in vertical or upward-flow designs. Figure 2.5 shows schematics of different types of sedimentation tanks used.

2.6.4 Empirical Equations for Settling Velocity

The foundational model for particle settling is Stokes' Law, which describes the motion of a small, spherical particle through a viscous fluid under laminar flow conditions. While this model offers valuable insights into simple systems, real-world suspensions; such as those encountered in wastewater treatment, are far more complex, often comprising particles of varying sizes, shapes, and densities. As a result, numerous empirical and semi-empirical models have been developed to estimate settling velocities under these conditions more accurately. These models aim to account for critical factors such as hydrodynamic forces, particle–fluid interactions, and inter-particle effects, which significantly influence settling behavior in concentrated or heterogeneous suspensions. Together, these empirical formulations provide engineers with practical tools for designing and optimizing sedimentation tanks under various operating scenarios.

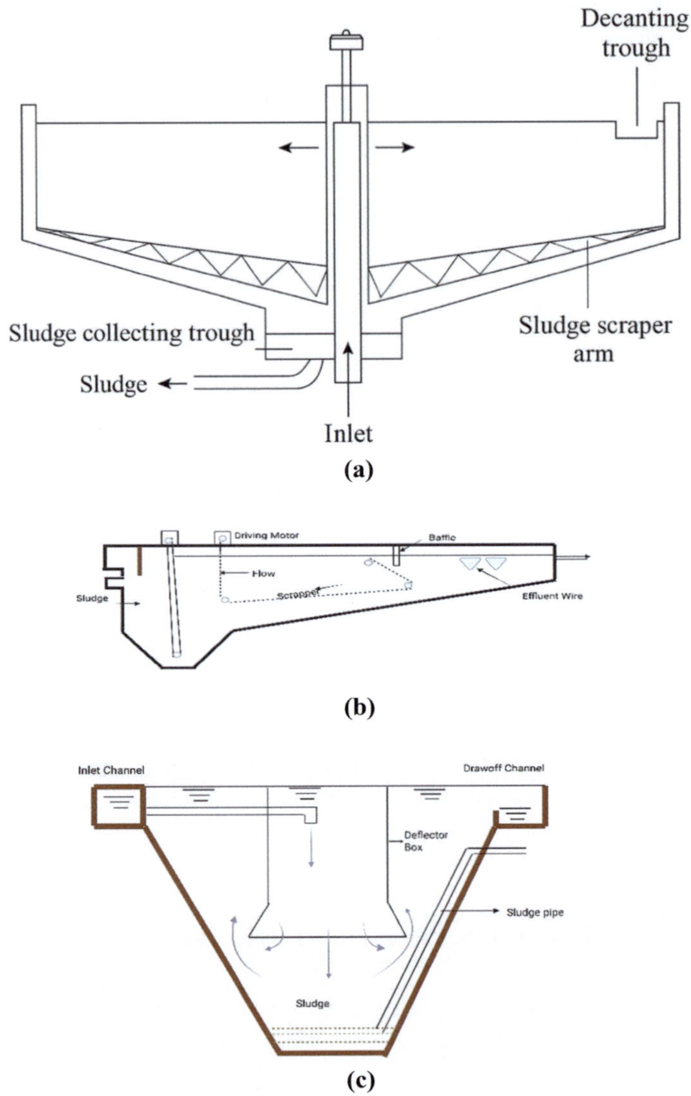

Fig. 2.5 Schematic of primary sedimentation tank **a** circular, **b** Rectangular (horizontal flow), **c** Hopper bottom tank

The following are critical empirical correlations widely used for estimating settling velocity during sedimentation.

2.6.4.1 Ferguson-Church Equation [35]

Ferguson-Church presented an equation for settling velocity as a function of grain diameter with known values of fluid viscosity and density. The equation can be applied to the entire range of viscous to turbulent conditions. The dimensional analysis derived the equation and converged to Stokes' law for small grains and a constant drag coefficient for large grains. It has two physically model parameters that can be adjusted for the shape effects of solid particles.

$$v_t = \frac{\rho g d^2}{C_1 \upsilon + \left(0.75 C_2 \rho g d^3\right)^{0.5}} \tag{2.15}$$

where C_1 and C_2 are model parameters.

The parameters C1 and C2 take values of 18 and 0.4 for smooth spheres, 18 and 1.0 for typical natural sands if sieve diameters are used, (20 and 1.1 if nominal diameters are used) respectively. A likely limit for very angular grains was suggested to be $C_1 = 24$, $C_2 = 1.2$.

2.6.4.2 Cheng Equation [36]

The Cheng equation is a generalized settling velocity model that is valid across the full range of Reynolds numbers, incorporating both laminar and turbulent flow effects. It is commonly used in sediment transport modeling.

$$v_t = \frac{\upsilon}{d}\left[\sqrt{25 + 1.2\left(\frac{d^3 g(\rho_s - \rho_l)}{\upsilon^2 \rho_l}\right)^{0.5}} - 5\right] \tag{2.16}$$

This is equation is obtained by assuming the following correlation for C_D and Re.

$$C_D = \left[\left(\frac{A}{Re}\right)^{\left(\frac{1}{n}\right)} + B^{\left(\frac{1}{n}\right)}\right]^n \tag{2.17}$$

where A and B are constants and n is exponent. This satisfies the two extreme conditions of Laminar and turbulent region. This equation implies C_D is inversely proportional to Re at low Reynolds numbers and becomes a constant at high Reynolds numbers.

2.6.4.3 Steinour Equation [37]

This empirical model accounts for non-spherical particles, often used in wastewater and industrial settling where particle shape deviates from ideal spheres. It includes a shape factor, improving upon Stokes' Law for irregular particles.

$$v_t = \frac{g d^2 (\rho_s - \rho_l)}{18\mu} S \tag{2.18}$$

where S: is shape factor which is 1 for spherical particles and < 1 for non-spherical particles.

2.6.4.4 Hallermeier Equation [38]

Developed for coastal sediment transport, this equation estimates the settling velocity of sands and silts under high-energy turbulent conditions such as surf zones. It applies to larger, natural sediment particles. This is essentially a correction to Stokes' Law using a drag correction.

$$v_t = \frac{1}{18}\left(\frac{(\rho_s - \rho_l)gd^2}{\mu}\right)\left[1 + 0.15\left(\frac{(\rho_s - \rho_l)gd^3}{\mu^2}\right)^{0.687}\right]^{-1} \tag{2.19}$$

2.6.4.5 Camp Equation [8]

This empirical relation is used primarily for design of sedimentation tanks handling flocculent particles, accounting for particle growth during settling. It assumes that floc formation enhances settling velocity with depth.

$$v_t = k.H^n \tag{2.20}$$

where H is the depth of settling zone, k and n are the empirical constants depending on floc characteristics

2.6.4.6 Richardson-Zaki Equation [39]

Widely used for hindered settling in suspensions, especially in thickeners and sludge. It models individual particle settling velocity reduction due to crowding effects at higher concentrations.

$$v_t = v_0(1 - C)^n \tag{2.21}$$

where:

v_0 is free settling velocity, C is solids volume fraction, and n is empirical exponent (typically 4.65 for small particles. The value of n is considered to vary with Re.

2.6.4.7 Michaels and Bolger Equation [40]

This equation is an extension of Richardson-Zaki for higher concentration suspensions. This model includes a term for compressibility, allowing it to be used for zone settling and compressive sludge modeling.

$$v_t = v_0(1 - C)^n \exp(- \propto C) \tag{2.22}$$

where: $\propto$ is the compressibility factor, and other terms are as described in Richardson-Zaki equations.

2.6.4.8 Batchelor Equation [41]

This equation is applicable for dilute suspensions. This model considers the hydrodynamic interactions between particles and is widely cited in theoretical fluid mechanics. However, it is less used in practical sedimentation tank design.

$$v_t = v_0(1 - 6.55C) \tag{2.23}$$

where: C is the volume fraction of solid spheres and v_0 is the velocity of a single sphere in unbounded fluid. The main assumption made for this model is that the centers of spheres in the dispersion take all positions with equal probability such that no two spheres overlap which is valid only when the spheres are identical.

2.6.4.9 Dietrich's Equation [42]

This is one of the most widely used empirical models in sediment transport and environmental engineering for calculating settling velocity of natural sediment. It incorporates particle shape, size, and density, and is valid across a wide range of Reynolds numbers. This model is based on non-dimensional quantities and was found to be satisfactorily over wide range of data published till then.

$$v_* = R_3 10^{R_1 + R_2} \tag{2.24}$$

where v_* is dimensionless settling velocity which is given by:

$$v_* = \frac{\rho {v_s}^3}{(\rho_s - \rho_l) g \nu} \tag{2.25}$$

The terms in Eq. 2.24 are obtained by fitted equations for size and density (R_1), shape (R_2) and roundness (R_3).

2.6.5 CFD-Based Tank Design

Conventional design methods for sedimentation tanks have historically relied on simplified assumptions such as weir overflow rates and detention times. These approaches often treat settling as a purely vertical process, ignoring the influence of complex hydrodynamics, turbulence, and interactions between solids and flow patterns [43]. Such simplifications fail to account for critical factors like atmospheric conditions, internal geometries, sludge properties, and flocculation [44]. As a result, these methods tend to rely on significant safety factors, potentially leading to oversized tanks or inefficient operation [45]. Moreover, they assume a constant settling velocity independent of solid concentration, despite clear evidence that hindered and compression settling are strongly concentration-dependent [46].

On the other hand, empirical equations derived based on certain assumptions often lack general applicability beyond the specific conditions under which they were developed. With the advent of Computational Fluid Dynamics (CFD), these limitations can now be addressed more precisely. CFD allows for a more realistic and comprehensive understanding of the flow and settling behavior within sedimentation tanks, making it a valuable tool for modern design and optimization.

CFD fundamentally changed how sedimentation tanks were studied and designed. Instead of relying on indirect indicators like surface overflow, CFD allowed engineers to simulate internal flow fields, visualize velocity profiles, turbulence zones, and particle trajectories. Using numerical techniques, CFD works by solving the Navier–Stokes equations (mass, momentum, and energy conservation laws). These partial differential equations govern fluid motion and particle behavior. Solving these equations allows engineers to model and visualize internal flow and optimize tank geometry and performance before construction [47].

2.6.5.1 CFD Tools and Software

Today, a variety of CFD platforms are available. Commonly used packages include: ANSYS Fluent, OpenFOAM, Star-CCM+, PHOENICS, Flowtherm, etc. These tools are accessible to researchers and can be run on standard desktops or laptops for moderately complex problems. However, high-fidelity simulations (e.g., involving 3D unsteady multiphase flows) may still require supercomputers due to high computational demands. CFD simulations not only complement laboratory experiments but also enable the study of systems and scenarios difficult to replicate in the lab, such as extreme wind effects, thermal currents, and rare hydrodynamic phenomena.

2.6.5.2 CFD Governing Equations

The CFD simulation of sedimentation tanks involves solving the following fundamental equations [46].

1. Momentum Equation (Navier–Stokes): This is a fundamental equation for any CFD simulation. It describes how velocity changes in space and time due to pressure gradients, viscous forces, gravity, and other body forces.
2. Continuity Equation (Mass Conservation): This equation describes the relation between fluid mass entering a control volume and the mass leaving it, accounting for accumulation.
3. Particulate Mass Conservation: This equation tracks the concentration of suspended solids and how they move and settle within the tank.
4. Settling Velocity Equation: Derived from Newton's and Stokes' laws, this describes how quickly a particle settles based on its size, density difference, and fluid viscosity.

Depending on tank geometry, these equations can be applied in Cartesian or cylindrical coordinate systems.

2.6.5.3 Two-Phase Flow Modeling

These equations are used to model either single-phase flow, i.e., fluid only, or multiphase flow, where both fluid and solids are considered. Although single-phase flow models are simpler and faster to execute, they are less accurate, making them less popular. Essentially, two-phase models are popular due to their accuracy. There are two main approaches for the two-phase model: Euler-Euler Model and Euler–Lagrange Model [48]. The Euler-Euler model treats both fluid and solids as interpenetrating continua. It is computationally efficient but may sacrifice accuracy in capturing particle-scale interactions. On the other hand, Euler–Lagrange Model tracks individual particles through the fluid field using a Lagrangian reference frame. It is more accurate for low solid concentrations (<10–12%). This model captures drag, lift, gravity, and other forces on each particle. Due to this, it is more computationally intensive than Euler-Euler. Bridgeman et al. [48] provide a detailed explanation of both models and their suitability in sedimentation applications.

2.6.5.4 Turbulence Modeling Techniques

In most sedimentation tanks, flow is turbulent rather than laminar. Accurately capturing this behavior is critical for predicting short-circuiting, dead zones, and mixing. There have been three primary turbulence modeling strategies that can be applied, and they are [49]:

(i) Reynolds-Averaged Navier–Stokes (RANS) Models: These models solve time-averaged Navier–Stokes equations by modeling turbulence effects. These are fast to solve and hence are widely used in industry. Examples are mixing length, k-ε, and k-ω model.
(ii) Reynolds Stress Model (RSM): These models provide better predictions for anisotropic turbulence and recirculating zones. However, they require more computational time than standard RANS models.
(iii) Large Eddy Simulation (LES): This model resolves large-scale eddies directly, thus modeling only smaller scales. These are more accurate and reliable than RANS. Gooarzi et al. [50] and Al-Sammarraee et al. [51] reported successful LES use in sedimentation tank CFD.

2.6.5.5 Baffle Configuration Analysis

The shift from empirical to CFD-based design has enhanced sedimentation tank performance analysis. CFD enables realistic simulations, supports custom designs, and identifies flow issues like short-circuiting. Though not a full replacement for experiments, CFD is a powerful complementary tool for optimizing designs and predicting fluid behavior in treatment systems. One of the key aspects of a sedimentation tank, which is analyzed and optimized using CFD-based design, is baffle configuration. Goula et al. [52] applied CFD to a full-scale potable water clarifier and discovered that introducing a vertical feed baffle reduced inlet recirculation zones, redirected particles downward, and improved solids removal efficiency from 90.4% to 98.6%. Goula et al. [53] demonstrated that even a 1 °C influent–tank temperature difference induces density currents that disrupt quiescent flow; strategically placed vertical feed baffles can suppress these plumes, restoring laminar conditions for improved solids retention. Saeedi et al. [54] used CFD to model multiple floor-level baffle configurations and found that fewer, lower-height baffles near the inlet effectively reduced circulation zones, thereby boosting sedimentation efficiency. Similarly, Shahrokhi et al. [55] paired CFD with Acoustic Doppler Current Profiler (ADCP) measurements to identify optimal lateral baffle placement. They concluded that positioning baffles to disrupt inlet jet patterns yields more uniform flow within the settling zone. Shahrokhi et al. [56] studied experimental and CFD methods to examine how the number of baffles affects flow patterns in sedimentation tanks. Results showed that increasing the baffles when optimally placed, reduces circulation zones and horizontal velocities, leading to more uniform flow and enhanced particle settling. The area occupied by the high value of kinetic energy also decreased by increasing the baffles. This produces a better area for the sedimentation process, thus proving that more baffles improve overall tank performance.

These studies demonstrate that CFD-based analysis significantly enhances the overall design of sedimentation tanks. By optimizing baffle number, height, and placement, CFD helps reduce short-circuiting, suppress turbulent recirculation, and control thermal density currents. These improvements result in more efficient settling conditions and better overall tank performance.

References

1. Riffat, R.; Husnain, T. *Fundamentals of Wastewater Treatment and Engineering*, Second edition.; CRC Press: Abingdon, Oxon ; Boca Raton, FL, 2022.
2. Deluise, F.; Wang, L. K.; Chang, S.-Y.; Hung, Y.-T. Screening and Comminution. *Physicochem. Treat. Process.* **2005**, 1–19.
3. Dold, P. L. Design and Control of Equalization Tanks. **1982**.
4. Mikola, A. The Effect of Flow Equalization and Low-Rate Prefermentation on the Activated Sludge Process and Biological Nutrient Removal. **2013**.
5. Foess, G. W.; Meenahan, J. G.; Blough, D. Evaluation of In-Line and Side-Line Flow Equalization Systems. *J. Water Pollut. Control Fed.* **1977**, 120–130.
6. Duan, J.; Gregory, J. Coagulation by Hydrolysing Metal Salts. *Adv. Colloid Interface Sci.* **2003**, *100*, 475–502.
7. Mattson, S. Cataphoresis and the Electrical Neutralization of Colloidal Material. *J. Phys. Chem.* **1928**, *32* (10), 1532–1552. https://doi.org/10.1021/j150292a011.
8. Camp, T. R. Velocity Gradients and Internal Work in Fluid Motion. *J Boston Soc Civ Eng* **1943**, *30*, 219–230.
9. Gregory, J.; O'Melia, C. R. Fundamentals of Flocculation. *Crit. Rev. Environ. Sci. Technol.* **1989**, *19* (3), 185–230.

10. Jiang, J.; Graham, N. Enhanced Coagulation Using Al/Fe (III) Coagulants: Effect of Coagulant Chemistry on the Removal of Colour-Causing NOM. *Environ. Technol.* **1996**, *17* (9), 937–950.
11. Zhang, M.; Xiao, F.; Xu, X.; Wang, D. Novel Ferromagnetic Nanoparticle Composited PACls and Their Coagulation Characteristics. *Water Res.* **2012**, *46* (1), 127–135.
12. Zhao, S.; Feng, C.; Huang, X.; Li, B.; Niu, J.; Shen, Z. Role of Uniform Pore Structure and High Positive Charges in the Arsenate Adsorption Performance of Al13-Modified Montmorillonite. *J. Hazard. Mater.* **2012**, *203*, 317–325.
13. Kristianto, H. The Potency of Indonesia Native Plants as Natural Coagulant: A Mini Review. *Water Conserv. Sci. Eng.* **2017**, *2* (2), 51–60.
14. Camacho, F. P.; Sousa, V. S.; Bergamasco, R.; Teixeira, M. R. The Use of Moringa Oleifera as a Natural Coagulant in Surface Water Treatment. *Chem. Eng. J.* **2017**, *313*, 226–237.
15. Zaidi, N.; Muda, K.; Abdul Rahman, M.; Sgawi, M.; Amran, A. Effectiveness of Local Waste Materials as Organic-Based Coagulant in Treating Water; IOP Publishing, 2019; Vol. 636, p 012007.
16. Anju, S.; Mophin-Kani, K. Exploring the Use of Orange Peel and Neem Leaf Powder as Alternative Coagulant in Treatment of Dairy Wastewater. *IJSER* **2016**, *7* (4), 238–244.
17. Teixeira, M. R.; Camacho, F. P.; Sousa, V. S.; Bergamasco, R. Green Technologies for Cyanobacteria and Natural Organic Matter Water Treatment Using Natural Based Products. *J. Clean. Prod.* **2017**, *162*, 484–490.
18. Bratby, J. *Coagulation and Flocculation in Water and Wastewater Treatment*; IWA publishing, 2016.
19. Ducoste, J. J.; Clark, M. M. The Influence of Tank Size and Impeller Geometry on Turbulent Flocculation: II. Model. *Environ. Eng. Sci.* **1998**, *15* (3), 225–235.
20. Ducoste, J. J.; Clark, M. M. Turbulence in Flocculators: Comparison of Measurements and CFD Simulations. *Am. Inst. Chem. Eng. AIChE J.* **1999**, *45* (2), 432.
21. Oliveira, D.; Teixeira, E. Hydrodynamic Characterization and Flocculation Process in Helically Coiled Tube Flocculators: An Evaluation through Streamlines. *Int. J. Environ. Sci. Technol.* **2017**, *14* (12), 2561–2574.
22. Cahyana, G. H.; Suwandhi, P.; Mulyani, T. Novel Helical or Coiled Flocculator for Turbidity Reduction in Drinking Water Treatment: A Performance Study; IOP Publishing, 2021; Vol. 623, p 012009.
23. Lin, L.; Li, R.; Li, Y.; Xu, J.; Li, X. Recovery of Organic Carbon and Phosphorus from Wastewater by Fe-Enhanced Primary Sedimentation and Sludge Fermentation. *Process Biochem.* **2017**, *54*, 135–139.
24. Shewa, W. A.; Dagnew, M. Revisiting Chemically Enhanced Primary Treatment of Wastewater: A Review. *Sustainability* **2020**, *12* (15), 5928.
25. Metcalf & Eddy; Abu-Orf, M.; Bowden, G.; Burton, F. L.; Pfrang, W.; Stensel, H. D.; Tchobanoglous, G.; Tsuchihashi, R.; AECOM (Firm). *Wastewater Engineering: Treatment and Resource Recovery*; McGraw Hill Education, 2014.
26. Taboada-Santos, A.; Rivadulla, E.; Paredes, L.; Carballa, M.; Romalde, J.; Lema, J. M. Comprehensive Comparison of Chemically Enhanced Primary Treatment and High-Rate Activated Sludge in Novel Wastewater Treatment Plant Configurations. *Water Res.* **2020**, *169*, 115258.
27. Sarparastzadeh, H.; Saeedi, M.; Naeimpoor, F.; Aminzadeh, B. Pretreatment of Municipal Wastewater by Enhanced Chemical Coagulation. **2007**.
28. Pasciucco, F.; Pasciucco, E.; Castagnoli, A.; Iannelli, R.; Pecorini, I. Comparing the Effects of Al-Based Coagulants in Waste Activated Sludge Anaerobic Digestion: Methane Yield, Kinetics and Sludge Implications. *Heliyon* **2024**, *10* (7), e29282. https://doi.org/10.1016/j.heliyon.2024.e29282.

29. Xu, G.; Zhang, W.; Li, G. Adsorbent Obtained from CEPT Sludge in Wastewater Chemically Enhanced Treatment. *Water Res.* **2005**, *39* (20), 5175–5185.
30. Yu, Z.; Jiang, Z.; Xu, X.; Xiao, F.; Zhang, Q.; Wang, C.; Zeng, M. Efficient Treatment of Leather Wastewater Using a Combination of CEPT and SBR with a Novel Flocculant. *J. Water Process Eng.* **2024**, *57*, 104575.
31. Fan, K.; Zhu, C.; Chen, G.; Zhu, X.; Wang, J.; Liu, F. Experimental Study on Surface Optimization of Dosage in Magnetic Coagulation Process for Enhancing Primary Treatment of Urban Sewage. *Desalination Water Treat.* **2024**, *320*, 100784.
32. Amoohadi, V.; Pasalari, H.; Esrafili, A.; Gholami, M.; Farzadkia, M. A Comparative Study on Polyaluminum Chloride (PACl) and Moringa Oleifera (MO) Chemically Enhanced Primary Treatment (CEPT) in Enhanced Biogas Production: Anaerobic Digestion Performance and the Gompertz Model. *RSC Adv.* **2023**, *13* (25), 17121–17129.
33. Gidstedt, S.; Betsholtz, A.; Cimbritz, M.; Davidsson, Å.; Hagman, M.; Karlsson, S.; Takman, M.; Svahn, O.; Micolucci, F. Chemically Enhanced Primary Treatment, Microsieving, Direct Membrane Filtration and GAC Filtration of Municipal Wastewater: A Pilot-Scale Study. *Environ. Technol.* **2024**, *45* (1), 28–39.
34. McCabe, W. L.; Smith, J. C.; Harriott, P. *Unit Operations of Chemical Engineering*; McGraw-hill, 1993.
35. Ferguson, R.; Church, M. A Simple Universal Equation for Grain Settling Velocity. *J. Sediment. Res.* **2004**, *74* (6), 933–937.
36. Cheng, N.-S. Simplified Settling Velocity Formula for Sediment Particle. *J. Hydraul. Eng.* **1997**, *123* (2), 149–152.
37. Steinour, H. H. Rate of Sedimentation. *Ind. Eng. Chem.* **1944**, *36* (9), 840–847.
38. Hallermeier, R. J. Terminal Settling Velocity of Commonly Occurring Sand Grains. *Sedimentology* **1981**, *28* (6), 859–865.
39. Richardson, J.; Zaki, W. The Sedimentation of a Suspension of Uniform Spheres under Conditions of Viscous Flow. *Chem. Eng. Sci.* **1954**, *3* (2), 65–73.
40. Michaels, A. S.; Bolger, J. C. Settling Rates and Sediment Volumes of Flocculated Kaolin Suspensions. *Ind. Eng. Chem. Fundam.* **1962**, *1* (1), 24–33.
41. Batchelor, G. Sedimentation in a Dilute Dispersion of Spheres. *J. Fluid Mech.* **1972**, *52* (2), 245–268.
42. Dietrich, W. E. Settling Velocity of Natural Particles. *Water Resour. Res.* **1982**, *18* (6), 1615–1626.
43. Hirom, K.; Devi, T. T. Application of Computational Fluid Dynamics in Sedimentation Tank Design and Its Recent Developments: A Review. *Water. Air. Soil Pollut.* **2022**, *233* (1). https://doi.org/10.1007/s11270-021-05458-9.
44. Stamou, A. I.; Theodoridis, G.; Xanthopoulos, K. Design of Secondary Settling Tanks Using a CFD Model. *J. Environ. Eng.* **2009**, *135* (7), 551–561. https://doi.org/10.1061/(asce)0733-9372(2009)135:7(551).
45. DeVantier, B. A.; Larock, B. E. Modeling Sediment-Induced Density Currents in Sedimentation Basins. *J. Hydraul. Eng.* **1987**, *113* (1), 80–94. https://doi.org/10.1061/(asce)0733-9429(1987)113:1(80).
46. Griborio, A. G. *Secondary Clarifier Modeling: A Multi-Process Approach*; University of New Orleans, 2004.
47. Anderson, J. D.; Wendt, J. *Computational Fluid Dynamics*; Springer, 1995; Vol. 206.
48. Bridgeman, J.; Jefferson, B.; Parsons, S. A. Computational Fluid Dynamics Modelling of Flocculation in Water Treatment: A Review. *Eng. Appl. Comput. Fluid Mech.* **2009**, *3* (2), 220–241.

49. Raeesh, M.; Devi, T. T.; Hirom, K. Recent Developments on Application of Different Turbulence and Multiphase Models in Sedimentation Tank Modeling—a Review. *Water. Air. Soil Pollut.* **2023**, *234* (1), 5.
50. Goodarzi, D.; Lari, K. S.; Alighardashi, A. A Large Eddy Simulation Study to Assess Low-Speed Wind and Baffle Orientation Effects in a Water Treatment Sedimentation Basin. *Water Sci. Technol.* **2018**, *2017* (2), 412–421.
51. Al-Sammarraee, M.; Chan, A.; Salim, S. M.; Mahabaleswar, U. Large-Eddy Simulations of Particle Sedimentation in a Longitudinal Sedimentation Basin of a Water Treatment Plant. Part I: Particle Settling Performance. *Chem. Eng. J.* **2009**, *152* (2–3), 307–314.
52. Goula, A. M.; Kostoglou, M.; Karapantsios, T. D.; Zouboulis, A. I. A CFD Methodology for the Design of Sedimentation Tanks in Potable Water Treatment: Case Study: The Influence of a Feed Flow Control Baffle. *Chem. Eng. J.* **2008**, *140* (1–3), 110–121.
53. Goula, A. M.; Kostoglou, M.; Karapantsios, T. D.; Zouboulis, A. I. The Effect of Influent Temperature Variations in a Sedimentation Tank for Potable Water Treatment—A Computational Fluid Dynamics Study. *Water Res.* **2008**, *42* (13), 3405–3414.
54. Saeedi, E.; Behnamtalab, E.; Salehi Neyshabouri, S. Numerical Simulation of Baffle Effect on the Performance of Sedimentation Basin. *Water Environ. J.* **2020**, *34* (2), 212–222.
55. Shahrokhi, M.; Rostami, F.; Said, M. A. M. Numerical Modeling of Baffle Location Effects on the Flow Pattern of Primary Sedimentation Tanks. *Appl. Math. Model.* **2013**, *37* (6), 4486–4496.
56. Shahrokhi, M.; Rostami, F.; Said, M. A. M.; Yazdi, S. R. S. The Effect of Number of Baffles on the Improvement Efficiency of Primary Sedimentation Tanks. *Appl. Math. Model.* **2012**, *36* (8), 3725–3735.

3 Secondary Biological Wastewater Treatment Methods

Abstract

This chapter reviews secondary biological wastewater treatment methods, emphasizing processes that reduce organic matter, suspended solids, and nutrients after primary treatment. It outlines suspended and attached growth systems, microbial metabolism pathways (aerobic, anaerobic, anoxic), and the principles of biokinetics for system design. The activated sludge process is discussed in detail, covering its configurations, operational parameters, advantages, and limitations. Biological nutrient removal through nitrification–denitrification and enhanced biological phosphorus removal is highlighted for achieving complete nutrient removal without chemicals. Anaerobic bioreactors such as UASB, CSTR, and AFBR are described, focusing on their energy efficiency and biogas production. The chapter also examines secondary clarification, membrane bioreactors, and their benefits, challenges, and advances, including hybrid systems. Finally, it addresses effluent disinfection methods like chlorination, ozonation, and UV emphasizing their mechanisms, operational considerations, and environmental implications for safe wastewater reuse or discharge.

Secondary treatment is a biological process aimed at significantly reducing the organic content of wastewater that remains after primary treatment. This stage focuses on the removal of dissolved and colloidal biodegradable organic matter (or BOD) and further reduction of suspended solids. At the heart of secondary treatment are microbial communities that metabolize organic pollutants into more stable forms such as carbon dioxide, water, and biomass. These processes occur in aerated or anoxic environments, depending on the specific treatment strategy employed. The main objectives of secondary treatment

A. H. Bari et al., *Wastewater Treatment*, Synthesis Lectures on Sustainable Development,
https://doi.org/10.1007/978-3-032-07585-7_3

include the removal of 85–95% of BOD and suspended solids, conversion of dissolved organics into biomass, and biological stabilization of organic matter. This process also prepares treated effluent for the safe discharge or additional treatment stages.

3.1 Principles of Biological Wastewater Treatment

Biological wastewater treatment is a cost-effective and environmentally favorable alternative because of the use of microorganisms to biodegrade toxic compounds. These microbes contribute to the cleanup of washed-out water through different systems, including activated sludge, biofilters, and wetlands. These processes are incredibly relevant regarding water conservation and reducing pollution [1]. Biological wastewater treatment method involves using various microorganisms, such as bacteria, to effectively degrade the organic pollutants in both industrial and municipal wastewaters. Biological treatment systems are generally classified based on how microorganisms interact with the wastewater. In suspended growth systems, such as activated sludge, sequencing batch reactors (SBR), and membrane bioreactors (MBRs), microorganisms are freely suspended in the wastewater and form flocs that degrade organic pollutants. In contrast, in attached growth systems like trickling filters, rotating biological contactors (RBCs), and moving bed biofilm reactors (MBBRs) microorganisms grow as biofilms attached to solid surfaces. The choice between these methods depends on factors like available space, energy requirements, desired effluent quality, and sludge production.

The metabolic activity of microorganisms dominates the biological wastewater treatment. These organisms get their energy and nourishment from the organic and inorganic substances found in wastewater. Through this, they transform the complex pollutants into simplified and less harmful ones like carbon dioxide, water, and biomass. Aerobic processes, i.e., the most frequent systems found are activated sludge, trickling filters, and SBRs systems, and anaerobic systems include up-flow anaerobic sludge blanket (UASB) reactors [2, 3]. The municipal wastewater is commonly treated using an aerobic process since this process removes the biodegradable organics, and nutrient and anaerobic process is mainly superior considering treatment of high-strength industrial wastewater and the retrieval of energy in the form of biogas [4]. This approach promotes the natural process of purification of the environment and overcoming serious environmental and health concerns, combined with performance improvement of treatment systems in terms of efficiency and effectiveness.

3.1.1 Microbial Metabolism

Microbial metabolism, or the breakdown of complex organic contaminants into simpler chemicals by certain microbial species, including bacteria and archaea, is the fundamental component of biological wastewater treatment. Depending on the availability of oxygen and source of alternative electron acceptors, the metabolic processes are divided into three groups: aerobic, anaerobic, and anoxic pathways. The breakdown not only assists in the elimination of the contaminants but also helps in reducing the BOD and COD in the wastewater [5].

A fundamental biological process in wastewater treatment is aerobic metabolism, in which bacteria oxidize structured contaminants by using molecular oxygen (O_2). This process primarily occurs in MBR, aerated lagoons, trickling filters, and activated sludge systems. Carbohydrates, proteins, and lipids are broken down by other microorganisms (mostly heterotrophic bacteria) to produce carbon dioxide (CO_2), water, and new biomass. In order to form Adenosine Triphosphate (ATP), the most significant type of energy currency in microbial cells, oxygen acts as the respiratory chain's final electron acceptor. These processes also absorb other metabolic byproducts, which are subsequently absorbed or mineralized [6, 7]. Biological wastewater treatment relies fundamentally on the metabolic activity of microorganisms that degrade organic pollutants. Microbial metabolism in such systems is broadly categorized into three pathways: aerobic, anaerobic, and anoxic.

3.1.1.1 Aerobic Metabolism

Within the context of the aerobic wastewater treatment systems, it is heterotrophic bacteria that break down organic matter to obtain energy and build up flocs in the activated sludge system. They are exergonic and use oxygen as a final electron acceptor in these metabolic reactions. A process of sedimentation segregates the resulting biomass. Major microbial genera in this process are *Pseudomonas, Bacillus,* and *Zoogloea*, which are very important in the consumption of organic materials found in the influent, thus making the treatment of the wastewater easy [2].

In aerobic processes, microorganisms use molecular oxygen (O_2) as the terminal electron acceptor. The carbonaceous organic matter is oxidized to carbon dioxide and water, with biomass production as a by-product. This process releases a significant amount of energy, supporting rapid microbial growth. Aerobic metabolism is commonly used in municipal wastewater treatment due to its effectiveness in reducing BOD and suspended solids.

3.1.1.2 Anaerobic Metabolism

Anaerobic wastewater treatment is a process where a micro community degrades organic waste without the presence of oxygen, and the final products are biogas, a gaseous mixture of CO_2 and methane. This method is an energy-efficient one, which is optimal when handling wastewaters from agricultural and municipal sources [3]. This converts the organic substance under the influence of *methanogenic archaea,* especially *Methanosarcina sp. and Methanotrix sp.,* into biogas consisting mainly of methane and CO_2 [8]. Anaerobic digestion has been a highly successful method of treating high-strength wastewater (HSWW), particularly that of industrial, dairy, and food origin. The most successful technology that is applicable presently is known as Anaerobic Membrane Reactor. It effectively eliminates the use of chemicals and reduces sludge formation because of mixing membrane filtration and biological processes, as they ensure 100 percent retention of biomass. The AMRs treatment of HSWW is a sustainable and energy-positive process because methane emissions are also minimized, and biogas becomes better harvested [9]. The breakdown of organic complex polymers (such as proteins, carbohydrates, and lipids) into more easily soluble organic molecules is the first phase, known as hydrolysis. Acidogenesis, which breaks down the exchanged products into volatile fatty acids (VFAs), ammonia, CO_2, and hydrogen, is then combined with this. Further conversion of VFAs in acetogenesis additionally produces hydrogen, CO_2, and acetic acid. These sequential microbial transformations are susceptible to environmental influences such as temperature, pH, and inhibiting substances [10].

Anaerobic processes occur in the absence of free oxygen, relying on alternative electron acceptors like sulfate, nitrate, or carbon dioxide. These processes are slower and yield less energy but produce valuable byproducts like methane and hydrogen sulfide. Anaerobic digestion is especially suitable for high-strength industrial wastewaters and for sludge stabilization, due to lower sludge production and energy recovery from biogas.

3.1.1.3 Anoxic Metabolism

Anoxic conditions, typically used in nitrogen removal processes, involve the use of nitrate or nitrite as electron acceptors in the absence of molecular oxygen. This is the basis of denitrification, where nitrate is reduced to nitrogen gas, thereby removing nitrogen from wastewater.

3.1.2 Biokinetics and Microbial Growth Models

Biokinetics is the study of the rates at which microorganisms metabolize substrates, and it plays a critical role in the design and operation of biological wastewater treatment systems. It encompasses mass balance, microbial population development, pollutant uptake, and environmental responses during treatment. This knowledge is essential for optimizing systems such as activated sludge, SBRs, biofilm-based systems, and MBR.

The foundation of biokinetics lies in microbial growth modeling, where the relationship between substrate concentration and biomass production is mathematically expressed. The most commonly applied model is the Monod model [11], which compares closely to enzyme kinetics. It relates the specific growth rate (μ) of microorganisms to substrate concentration (S) using two key parameters: maximum specific growth rate (μ_{max}) and half-saturation constant (K_S). The Monod equation is expressed as:

$$\mu = \frac{\mu_{max} S}{K_s + S} \tag{3.1}$$

where μ is the specific growth rate of the microorganisms,

μ_{max} is the maximum specific growth rate of the microorganisms,

S is the substrate concentration,

X is the biomass concentration and

K_S is the half-saturation or half-velocity constant (also known as substrate affinity constant).

This model enables engineers to estimate vital design parameters such as food-to-microorganism (F/M) ratio, hydraulic retention time (HRT), and sludge age by interpreting microbial response to nutrient levels. However, because real systems rarely operate at optimal conditions, the Monod model is often modified to include factors like decay or inhibition. Beyond growth, biokinetic models also account for endogenous decay (kd) and biomass yield (Y). The net biomass production is given by [13]:

$$\mu_{net} = Y\frac{kS}{K_s + S} - b \tag{3.2}$$

where μ_{net} is net specific biomass growth rate,

Y is synthesis yield coefficient,

b is the specific endogenous decay coefficient,

Under specific operational conditions, alternative kinetic models such as Contois, Tessier, and Moser may be more suitable than the traditional Monod model, particularly when addressing variations in biomass concentration or substrate availability [13]. To account for these complexities, advanced simulation tools like GPS-X, BioWin, and WEST incorporate a range of biokinetic models, including the widely adopted Activated Sludge Models (ASM1, ASM2, ASM3) [14]. These tools enable the simulation of critical biological processes such as nitrification, denitrification, and biological phosphorus removal, and are used extensively in the design, optimization, and troubleshooting of wastewater treatment plants.

3.1.3 Activated Sludge Microbiology

Activated sludge microbiology is the study of the complex microbial communities that drive the biological treatment processes in activated sludge systems. Such systems are often applied in wastewater treatment plants to remove organic and toxic substances, as well as nutrients. There is a need to thoroughly understand a wide range of microbial physiology and ecology in the activated sludge system to optimize process performance, stability, and regulate the effluent quality [15]. Microorganisms that are found in the ecology of the activated sludge are numerous and include bacteria, archaea, protozoa, fungi, and eukaryotes [16]. These are the most common and diverse, which include the bacteria that are important in organic materials decomposition and in the retrieval of nutrients. The most common phyla of bacteria are *Proteobacteria, Bacteroidetes, Chloroflexi,* and *Actinobacteria.* Nitrification is based on the growth of nitrifying organisms such as *Nitrosomonas* and *Nitrospira,* whereas phosphorus removal is mostly performed by polyphosphate-accumulating organisms (PAOs) and glycogen-accumulating organisms (GAOs) [17]. Some of the most common bacterial taxa in activated sludge systems include *Proteobacteria, Bacteroidetes, Acidobacteria, Firmicutes,* and nitrifying bacteria belonging to *Nitrospirae,* and many other microbial consortia. In urban WWTPs, very regularly, members of the *Proteobacteria* may constitute up to 60% of the abundance of the community and are essential to the nitrogen cycle and degradation of organic carbon [18]. New research on the activity of protozoan and metazoan predators focuses on the regulation of bacterial population growth by these predators, which can facilitate the elimination of sludge and floc formation. This predation does have the potential to reduce biomass growth down to virtually nothing, but increase the diversity and settlement character of the sludge. Further, bacteriophages have been shown to affect microbial ecology through lysis of rapidly growing bacteria, releasing nutrients required to assist community dynamics and sludge stability [19]. The microbial community structure changed greatly in laboratory-scale SBRs in the treatment of landfill leachate, and mainly due to enrichment of certain heterotrophic denitrifiers and alteration of sludge floc structure [20]. Likewise, cooking wastewater treatment showed seasonally varying genera such as *Thiobacillus, Cloacibacterium,* and *Pseudomonas* and functional redundancy, where removal performance remained constant across variations in community composition, with dissolved oxygen regarded as a major shaping parameter [21]. The activated sludge process is a cornerstone of secondary wastewater treatment, relying on a diverse and dynamic microbial community. Understanding the microbial ecology of this system is critical for optimizing treatment efficiency, preventing operational problems, and enabling advanced processes like nutrient removal.

3.1.4 Biological Nutrient Removal

Biological nutrient removal (BNR) is a necessity to ensure that excess nitrogen and phosphorus are eliminated in sewage before its discharge to the environment. Despite the nutrients being essential to the growth of plants, their excessive abundance may result in negative impacts on the environment, such as eutrophication in waters. Eutrophication is the process in which water bodies gets over enriched in nutrients, encouraging the growth of algae and aquatic plants which leads to water quality degradation and habitat loss. Although, phosphates are considered to be the primary source of eutrophication, nitrogenous wastes also contribute significantly to this problem [22]. Due to this biological nutrient removal is an essential as this process converts and oxidizes these nutrients utilizing naturally occurring microbial activities.

3.1.4.1 Nitrogen Removal

The two main biological processes by which nitrogen removal takes place through BNR are Nitrification and Denitrification [23].

(i) Nitration: The first process of nitritation is ammonia oxidation and is carried out in by two groups of organisms, ammonia-oxidizing bacteria (AOB) and ammonia-oxidizing archaea, which convert ammonium to nitrite.

$$NH_4^+ + 1.5\,O_2 \rightarrow NO_2^- + H_2O + 2H^+$$

In the second step the nitrite oxidizing bacteria oxidizes nitrite to nitrate.

$$NO_2^- + 0.5\,O_2 \rightarrow NO_3^-$$

The overall nitration reaction is given by:

$$NH_4^+ + 2\,O_2 \rightarrow NO_3^- + H_2O + 2H^+$$

Normally the first step of ammonia oxidation is the slow step, and hence little NO_2 accumulation is observed bioreactors.

(ii) Denitrification follows nitrification and occurs under anoxic conditions i.e. in the absence of dissolved oxygen, but in presence of nitrate. Denitrification is the conversion of nitrate to unavailable forms of nitrogen, mostly as nitrogen gas (N_2), with the aid of denitrifying bacteria, in anaerobic conditions. This is an anaerobic reaction that requires the availability of an organic source of carbon. Heterotrophic bacteria use organic carbon as the electron donor and nitrate (NO_3^-) as the electron acceptor, reducing it stepwise to nitrogen gas (N_2):

$$NO_3^- \rightarrow NO_2^- \rightarrow NO \rightarrow N_2O \rightarrow N_2 \uparrow$$

This gas is released to the atmosphere, completing nitrogen removal. Denitrification is facilitated by proper reactor sequencing: aerobic–anoxic zones, step-feed systems, or internal recycling from aerobic to anoxic reactors.

3.1.4.2 Phosphorus Removal

Phosphorus promotes the growth of algae and toxic cyanobacteria, contributing to eutrophication, and must be removed from wastewater to protect aquatic ecosystems [24]. Increasingly strict regulations have driven wider adoption of P removal. Enhanced biological phosphorus removal (EBPR) eliminates P without chemical addition by cycling activated sludge through anaerobic and aerobic phases [25]. Biological nutrient removal (BNR) combines EBPR with nitrogen removal. Polyphosphate accumulating organisms (PAOs) are key to phosphorus removal, storing it as intracellular polyphosphate that is removed with waste activated sludge (WAS). EBPR involves cycling biomass between anaerobic and aerobic environments. Unlike most microbes, PAOs takes up VFAs anaerobically and store them as poly-β-hydroxyalkanoates (PHAs), using energy from polyphosphate cleavage and phosphate release. In the aerobic zone, PAOs use stored PHAs for growth and energy, taking up large amounts of phosphate from the liquid and storing it intracellularly as polyphosphate granules.

This results in a net removal of phosphorus when sludge containing polyphosphate-rich biomass is wasted.

$$\text{Anaerobic : VFA uptake} \rightarrow \text{PHA storage} \rightarrow \text{Poly} - \text{P breakdown} \rightarrow \text{PO}_4^{3-}\text{release}$$

$$\text{Aerobic : PHA oxidation} \rightarrow \text{ATP generation} \rightarrow \text{Poly} - \text{P synthesis} \rightarrow \text{PO}_4^{3-}\text{uptake}$$

Effective EBPR requires a readily biodegradable carbon source, proper anaerobic–aerobic cycling, and control of SRT and pH. Its performance may be hindered by GAOs, which compete with PAOs without removing phosphorus. Combined with nitrification–denitrification, EBPR enables complete biological nutrient removal without chemicals and is widely used in plants discharging to nutrient-sensitive waters [3].

3.2 Activated Sludge Process

The Activated Sludge Process (ASP) is the most widely used biological treatment technology for the removal of organic pollutants and nutrients from wastewater. With a history of over a century since its development [26], ASP has evolved into a robust and reliable treatment method. It involves the suspension of microbial biomass in an aerated tank, where the mixed liquor degrades organic matter, followed by solid–liquid separation in a secondary clarifier. The activated sludge process is the most established and widely used biological treatment method for municipal and industrial wastewater. After primary treatment, which removes suspended solids, the wastewater undergoes biological treatment in a system consisting of an aeration tank followed by a secondary clarifier. The aeration tank functions as a completely mixed or sometimes plug flow bioreactor, where a specific concentration of biomass (MLSS or MLVSS) is maintained. Adequate dissolved oxygen levels are ensured to support the microbial degradation of soluble organic matter.

3.2.1 Process Overview

In a conventional ASP, the primary-treated wastewater enters an aeration tank, where it is mixed with RAS from the secondary clarifier. This MLSS is supplied with air to maintain dissolved oxygen levels of about 1.5–2.5 mg/L, which supports microbial growth and activity. In the aeration tank, microorganisms break down organic matter, reducing the carbonaceous BOD, and, if the SRT is long enough, carry out nitrification. The bacteria also produce EPS that help form flocs. The mixed liquor then flows to a secondary clarifier, where these flocs settle. Part of the settled biomass is returned to the aeration tank as RAS, while the rest is removed as WAS to control biomass levels and maintain the desired sludge age. Important operating parameters include the HRT, typically 4–8 h, and SRT, which ranges from about 5 days for carbon removal to around 15 days when nutrient removal is also required. The F/M ratio is adjusted to balance the organic load with the available biomass. When operated under well-controlled conditions, the ASP can remove 85–95% of BOD, COD, and suspended solids.

3.2.2 Variants of the Activated Sludge Process

Over time, several modifications of the conventional ASP have been developed to address varying wastewater characteristics, space limitations, operational preferences, and treatment goals. These variants alter hydraulic flow patterns, aeration strategies, or sludge handling methods to achieve specific performance outcomes.

(i) Extended Aeration (EA) process: In this process, a low-rate variation of ASP characterized by long HRT (18–30 h) and sludge SRT (20–30 days). This extended treatment ensures complete stabilization of organics within the aeration tank and results in minimal sludge production due to extensive endogenous respiration. EA systems are well-suited for small communities, resort areas, and remote installations, offering operational simplicity and low sludge yield, though they require higher energy input due to continuous aeration.

(ii) Contact Stabilization (CS) process: This process involves an initial short-term aeration phase in a contact tank, where incoming wastewater is mixed with return sludge for about 1–2 h. The mixed liquor is then clarified, and the settled biomass is transferred to a separate stabilization tank for extended aeration (4–6 h) to stabilize absorbed organics. By separating the contact and stabilization phases, this method allows for rapid BOD removal, shorter HRTs, and reduced aeration tank volumes, making it particularly suitable for treating high-strength or industrial wastewaters and handling variable or shock loads effectively.

(iii) Step-Feed process: This process on the other hand, divides the aeration tank into multiple zones, introducing influent wastewater and return sludge at several points along the treatment path. Each compartment receives its portion of the flow, with aeration provided either independently or in a shared manner. This configuration promotes improved BOD removal, stabilizes oxygen demand, and enhances denitrification through the creation of gradient conditions. Step-Feed systems are advantageous for better process control, prevention of front-end overloading, and efficient nutrient removal under high-load variations.

Overall, the choice of ASP variant depends on influent characteristics, plant capacity, and regulatory requirements, with each approach designed to meet specific objectives like whether minimizing sludge yield, maximizing throughput, or enhancing nutrient removal.

3.2.3 Types of Reactors

Over the time, various ASP configurations have evolved to meet changing treatment demands. These developments were driven by the need for higher-quality effluents, effective nutrient removal, and improved understanding of microbial processes and their fundamentals. Advances in equipment, materials, electronics, and process control have further enabled innovation. Similarly, the ongoing pressure to reduce capital, operating, and energy costs has encouraged the adoption of diverse reactor designs for ASP. The configuration of reactors in the ASP significantly affects treatment efficiency, energy usage and flexibility. Based on flow regime several types of biological reactor configurations are used in practice most common are plug flow, complete-mix and sequential batch process (SBRs) (Fig. 3.1).

3.2.3.1 Completely Mixed Activated Sludge Reactor (CMAS)

The conventional ASP in its basic form, employes a stirred tank bioreactors in a suspended-growth system, this is referred as the CMAS reactor. It was developed partly to dilute toxic industrial discharges but became prevalent in the 1970s and 1980s. In CMAS systems, contents are thoroughly mixed, and ideally the concentrations of the soluble substances (i.e., COD, BOD, NH_4^-, N) should be uniform throughout the tank. The process uses a large aerobic tank seeded with recycled microbial sludge, where air is sparged from the bottom to maintain sufficient dissolved oxygen for microbial metabolism. Alongside carbon removal, nitrification occurs in the aerobic tank, while denitrification place in a separate anoxic reactor without aeration. Over the time, multiple CSTR-based configurations have evolved to improve performance. The Bardenpho process uses four CSTRs in series (i) pre-denitrification, (ii) aerobic, (ii) post-denitrification, and (iv) final aerobic stage. Step aeration divides the aerobic tank into several compartments, each aerated separately, improving BOD destruction and reducing dead zones. Similarly, series–parallel arrangements incorporate step feeding and aeration, where each compartment receives both fresh influent and partially treated wastewater, enhancing substrate–microorganism contact and improving overall efficiency particularly in large-scale plants. These CSTR-based designs remain versatile and adaptable, forming the basis for numerous modern ASP configurations.

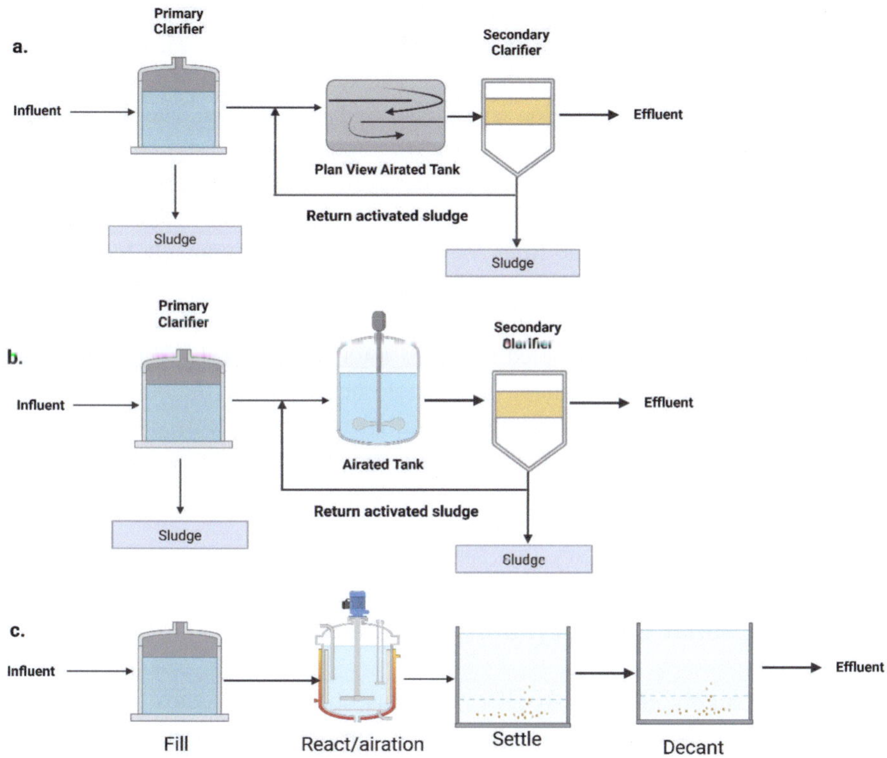

Fig. 3.1 Various reactor configuration of activated sludge processes: **a** flow diagram of plug flow process **b** flow diagram of complete-mix process **c** SBRs process

3.2.3.2 Plug Flow Reactor (PFR)

The conventional PFR application to ASP originated from narrow aeration tank designs (8–10 m wide, 5–6 m deep, and often over 100 m long) using sidewall coarse bubble aeration. Settled wastewater and RAS enter at the front and pass through multiple channels before reaching the secondary clarifier. Oxygen demand is highest at the inlet and tapers toward the outlet, leading to the concept of "tapered aeration," with higher aeration rates at the front and lower at the end. True plug flow is rarely achieved due to back mixing from aeration, although baffle walls can create a series of reactors in sequence. Widely used from the 1920s to the 1970s, plug flow ASPs became less suitable when industrial discharges increased in the late 1960s, as toxic wastes adversely affected process stability. Compared to complete mix ASPs, plug flow reactors have long, narrow basins with concentration gradients of soluble and suspended solids along their length, whereas complete mix systems maintain uniform concentrations throughout. In practice, many PFR configuration's function more like a series of complete-mix reactors due to hydrodynamic mixing effects [27].

3.2.3.3 Sequencing Batch Reactor (SBR)

The SBR is a type of fill-and-draw activated sludge system used for wastewater treatment. It performs all key steps of aeration, settling, and decanting within a single complete-mix reactor, eliminating the need for separate secondary clarifiers. The process operates in cycles, typically including fill, react (aeration), settle, decant, and sometimes an idle phase to accommodate flow variations. A typical cycle might consist of 3 h of filling, 2 h of aeration, 0.5 h of settling, and 0.5 h of supernatant withdrawal[13]. SBRs became more widely used in the late 1970s due to the development of programmable logic controllers (PLCs), level sensors, and automated valves, making them efficient and cost-effective, especially for small communities and industrial facilities with intermittent flow. More recently, SBRs have also been adopted for larger municipal systems [28]. Multiple basins are typically used to ensure continuous treatment one fills while the other completes the remaining steps. Aeration is provided through diffusers or jet systems, and sludge wasting generally occurs during aeration. The HRT ranges from 18 to 30 h, depending on flow and tank volume. SBRs also support anoxic or anaerobic processes during the fill phase by enabling separate mixing without aeration.

3.2.4 Operational Considerations and Challenges

The ASP, while highly effective, requires rigorous operational control to maintain consistent performance. Several challenges can arise during routine operation, necessitating timely diagnosis and corrective action. Sludge bulking is a common issue, caused by excessive growth of filamentous bacteria such as *Sphaerotilus natans* or *Microthrix parvicella*, which hinder floc formation and settling in the secondary clarifier [29]. This leads to poor settleability, high effluent turbidity, and solids carryover. Control measures include adjusting the food-to-microorganism (F/M) ratio, applying selectors or chlorination in severe cases, and increasing dissolved oxygen (DO) levels to discourage filament growth. Foam formation is a common issue in ASP which is often caused due to filamentous bacteria like *Nocardia* and *Gordonia* [30]. This potentially causes effluent quality deterioration, odors, increased maintenance time, hazardous conditions from overflow, and higher operational costs. Control measures include reducing aeration, lowering grease and oil in influent, chlorinating RAS or surfaces, using selectors, adjusting SRT, and applying antifoams or polymers. Nevertheless, rapid and accurate identification of the foam origins is the first and essential step for effective foam management.

DO control is critical, with optimal levels around 1.5–2.5 mg/L; low DO can result in incomplete BOD removal or inhibit nitrification, while excessive DO wastes energy. Due to changes in influent load, temperature, microbial activity, or aeration efficiency, DO levels can fluctuate. Advanced DO sensors and control systems like model predictive control can be effectively used to regulate dissolved oxygen levels in wastewater treatment plants enabling real-time regulation [31]. Maintaining appropriate SRT (~2–5 days for conversion of organics, ~ 3–18 days for complete nitrification, and ~ 2–4 days for phosphorous removal) [12] and F/M ratio is essential for stable operation.

Other key considerations include maintaining optimal MLSS and MLVSS levels and their organic fraction respectively to decide on sludge recycling, aeration demand, and process control. RAS flow rates generally range from 25–100% of influent flow [32], while WAS removal is critical for controlling sludge age and preventing bulking or foaming. Startup requires seeding with acclimated biomass and gradual organic loading, while environmental factors such as temperature, pH (6.5–8.5), and alkalinity (>50 mg/L as $CaCO_3$) must be controlled to sustain microbial activity. Automation and process monitoring of parameters like DO, pH, MLSS, and SVI support efficiency, while optimizing aeration is vital, as it accounts for up to 60% of a plant's energy demand. Ultimately, the success of ASP operation hinges on maintaining a balance between microbial activity, substrate availability, oxygen supply, and biomass control, supported by robust design and careful process management.

3.2.5 Advantages and Limitations of ASP

The ASP is the most widely adopted biological wastewater treatment technology due to its high efficiency, design flexibility, and scalability [12]. Under optimized conditions, ASP can remove 85–95% of BOD, COD, and suspended solids, and with process enhancements such as anoxic zones or step-feed arrangements, it is also effective in biological nutrient removal, eliminating the need for chemical dosing and reducing sludge generation. Its mature status in the field means there is a wealth of design standards, modeling tools (e.g., ASM1), and operational knowledge, making it a dependable choice for both small and large-scale facilities. Moreover, customizable configurations like extended aeration, SBRs, oxidation ditches, and MBRs make ASP adaptable to diverse treatment goals and space constraints.

Despite these strengths, ASP has some limitations, particularly in terms of operational demands and energy use. One of the major drawbacks is its high energy demand, especially due to continuous aeration, which can account for 50–60% of a plant's total energy consumption [13]. In the aeration tank, biodegradable organics are oxidized to inorganic compounds. Theoretically unlimited aeration time and an abundant supply of microorganisms, is needed to achieve complete oxidation. However, economic constraints prevent providing enough time for full oxidation, even in extended aeration processes. Also maintaining a constant population of these microorganisms is not always practical [33]. The production of WAS is significant and its management and disposal or treatment requires further thickening, stabilization, and dewatering. All of these processes add to operational complexity and incurs substantial costs [34]. ASP systems can be sensitive to toxic or industrial shocks, which may inhibit microbial activity and upset the performance of system [35]. Additionally, the system demands careful process control, requiring real-time monitoring of parameters like sludge age, DO, and F/M ratio to maintain stable operation [36]. Operational challenges such as sludge bulking and foaming, often due to filamentous bacteria like *Nocardia* or *Microthrix*, can compromise settling and effluent quality. Lastly, conventional ASP configurations occupy large space, which limits their feasibility in space-constrained urban settings unless compact designs such as MBRs are adopted.

In summary, while ASP remains a workhorse technology in biological wastewater treatment, its full potential depends on careful design, automation, and active process control to mitigate its inherent limitations.

3.3 Attached Growth Process

Attached growth process is a biological treatment process in which microorganisms that convert organic matter or other substances in wastewater into gases and biomass are attached to an inert medium such as rocks, slag, or specially designed plastic or ceramic materials. These processes can be classified into five general types [12], schematic of few of them is given in Fig. 3.2:

(i) Non-submerged attached growth systems, these systems, like trickling filters, use fixed media (rock or plastic) exposed to air, where wastewater is distributed over the media and treated as it flows through the biofilm.
(ii) Partially submerged systems. Reactors like Rotating Biological Contactors (RBCs) involve rotating discs partially submerged in wastewater, allowing biofilm to alternately get in contact with air and wastewater for treatment.
(iii) Sequential non-submerged attached growth combined with activated sludge. This system combines trickling filters with activated sludge to improve treatment efficiency and effluent quality by using each process in sequence.
(iv) Submerged attached growth systems. These include fixed or moving media submerged in aerated reactors, allowing compact treatment with enhanced biofilm activity and reduced settling issues.
(v) Activated sludge processes incorporating biofilm carriers. These systems incorporate biofilm-supporting media within activated sludge tanks to boost treatment capacity and improve biological performance.

Despite the variations, these systems share several key characteristics. First, microbial biomass grows on an inert medium such as plastic, ceramic, or stone. As the biofilm matures, excess solids periodically slough off and are removed either by sedimentation or through backwashing of the media. Oxygenis supplied differently depending on the system design; either by air circulation through void spaces in non-submerged systems or by

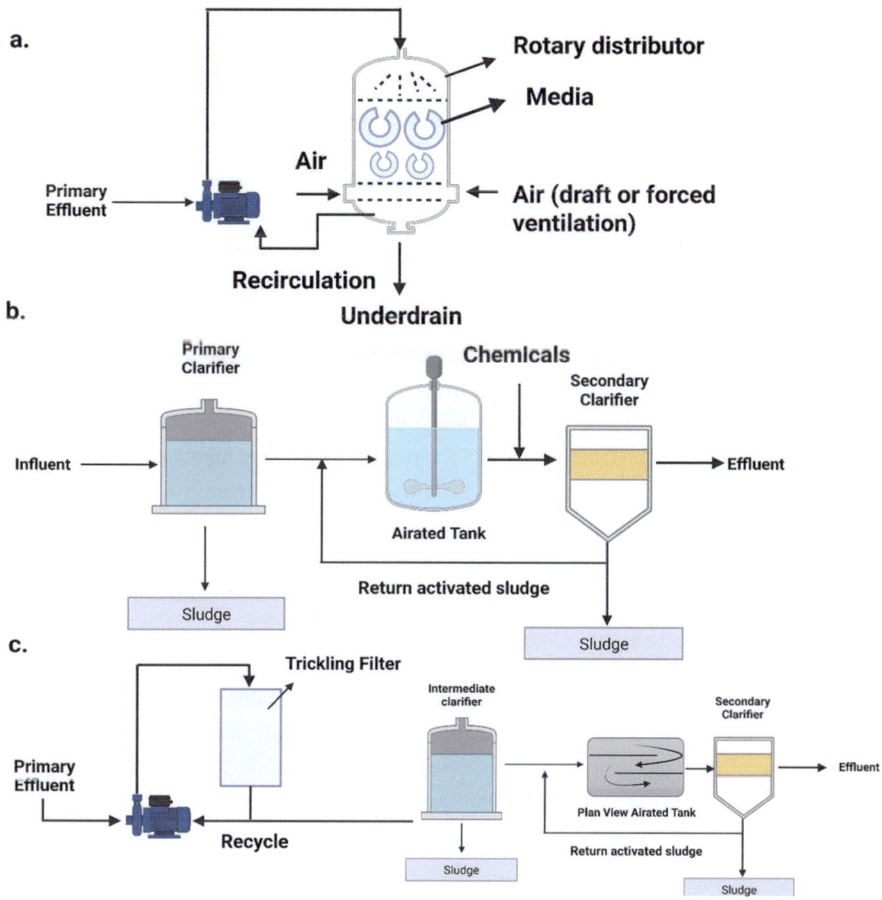

Fig. 3.2 Different types of attached growth process **a** tower type trickling filter, **b** MBBR system for BOD removal and phosphorus removal by chemical precipitation. **c** An integrated fixed film activated sludge

air sparging in submerged configurations. In some cases, oxygen is added to recycle flows, particularly in fluidized bed reactors. Proper distribution and contact between the wastewater and the media surface are also necessary to ensure effective treatment. Additionally, systems require underdrains or collection mechanisms to capture the treated effluent. A key characteristic of attached growth processes, unlike activated sludge systems, is that their performance is often limited by diffusion within the biofilm [39]. Substrate removal and biological reactions occur through layers of the biofilm, making removal rates dependent on the diffusion of oxygen and nutrients. This can lead to the formation of both aerobic and anaerobic zones within the biofilm, allowing simultaneous nitrification and denitrification even under aerobic bulk liquid conditions.

3.4 Anaerobic Bioreactors

Anaerobic reactors are widely used in wastewater treatment for their ability to degrade organic matter in the absence of oxygen, producing biogas as a valuable by-product. Anaerobic reactors offer several advantages over aerobic systems, making them an attractive option in certain wastewater treatment applications [12]. They require less energy due to the absence of aeration, and generate lower quantities of biological sludge. Additionally, they consume fewer nutrients, such as nitrogen and phosphorus, and can produce biogas as a potential renewable energy source. Their compact reactor size, reduced air pollution, and ability to resume operation quickly after extended idle periods add to their operational benefits. Anaerobic systems also function well as pretreatment processes and are associated with a lower carbon footprint.

Common types of anaerobic reactors used in wastewater treatment include [12] the Upflow Anaerobic Sludge Blanket (UASB) reactor, Continuous Stirred Tank Reactor (CSTR), Anaerobic Filter (AF), Anaerobic Fluidized Bed Reactor (AFBR), Expanded Granular Sludge Bed (EGSB) reactor, and the Anaerobic Sequencing Batch Reactor (ASBR). Among these designs, the most important types include the UASB reactor, CSTR, and AFBR, each with specific structural and operational feature. Figure 3.3 shows schematic of UASB, CSTR and AFBR.

The UASB reactor is one of the most popular systems for municipal and industrial wastewater treatment. It consists of a vertical cylindrical tank where wastewater flows upward through a dense sludge bed composed of anaerobic granules. Gas–solid-liquid separation is achieved using a three-phase separator located at the top of the reactor. Its main advantages include low energy consumption, biomass retention without mechanical mixing, and good resistance to load fluctuations.

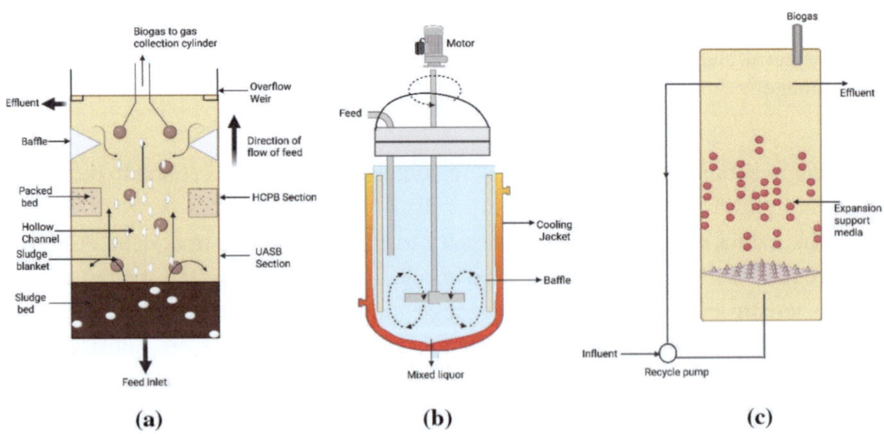

Fig. 3.3 Common types of anaerobic bioreactors **a** UASB reactor **b** CSTR **c** AFBR

The CSTR is a completely mixed system, typically used for high-solid content waste such as sewage sludge or manure. It includes a heated, insulated tank equipped with mechanical or gas mixing to ensure uniform substrate distribution and temperature. Although it requires higher energy input, its advantages lie in its simplicity, operational flexibility, and ability to handle a wide variety of substrates.

The AFBR features inert or synthetic media in suspension, which provides a large surface area for biomass attachment. The upward flow of wastewater keeps the bed fluidized, enhancing mass transfer and microbial contact. This reactor is particularly advantageous for treating high-strength or inhibitory wastewaters, offering compact design and high treatment efficiency. MBRs, which offer advanced biological treatment with membrane-based separation, are also increasingly used in aerobic digestion and are discussed in detail later in the chapter.

Anaerobic reactors come with certain limitations. They typically have a longer startup time as microbial biomass develops more slowly. These systems may also require alkalinity supplementation, and in many cases, post-treatment using aerobic processes is needed to meet effluent discharge standards. Furthermore, biological nitrogen and phosphorus removal is not achievable in anaerobic systems. Their performance is more sensitive to low temperatures, and they are more vulnerable to process upsets due to toxic shocks or variable influent loads. Lastly, there is potential for odor issues and concerns regarding the corrosiveness of biogas produced during digestion.

3.5 Secondary Clarification

Most biological treatment processes produce excess biomass through the conversion of waste carbon to new cells. As such, before the final treatment steps, such as disinfection or nutrient removal, solids must be separated from the secondary treatment effluents. This is usually by settling, but membranes are also employed which is the basis of Membrane bio reactors. The separated solids are either recycled back to the head of the process train or sent to digesters for solids reduction and processing, depending on the type of the digester system.

Secondary clarifiers are essential components of the biological wastewater treatment process, serving to separate treated effluent from biological solids following aeration. After wastewater passes through the aeration tank where microorganisms consume dissolved organic matter, the resulting mixed liquor, rich in microbial flocs, flows into the secondary clarifier. Here, gravity settling allows the clarified liquid to overflow for discharge or further treatment, while the settled sludge is either recycled back to the aeration tank as RAS or removed as WAS to maintain biomass balance.

Unlike primary clarifiers, which remove inorganic and settleable solids from raw wastewater, secondary clarifiers deal with biologically active solids and require more precise design to accommodate zone settling and maintain floc integrity. Their function is

Table 3.1 Comparison between primary and secondary clarifier

Characteristic	Primary clarifier	Secondary clarifier
Function	Pre-biological solids removal	Post-biological solids separation
Solids type	Inert and organic settleable solids	Biological floc (activated sludge)
Settling type	Type I & II	Type III (Zone), Type IV (Compression)
Sludge usage	Sent to digester	RAS & WAS split for reuse/removal
Detention time	1.5–2.5 h	4–6 h
equipment complexity	Simple	More complex (feed well, baffles)

tightly coupled with the biological stage, and their performance directly impacts effluent quality, sludge age, and process stability. To minimize hydraulic short-circuiting and sludge blanket disturbances, secondary clarifiers often include features like central feed wells, energy-dissipating inlets, and effluent weirs strategically placed for uniform withdrawal [38]. A brief comparison between primary and secondary clarifier is given in Table 3.1.

Recent studies have focused on optimizing secondary clarifier performance under variable flow and load conditions. CFD analyses have been used to redesign inlet geometry and baffle placement to reduce sludge bulking and enhance solids retention [39]. Advanced modeling approaches now consider flocculation kinetics, settling velocity distribution, and hydraulic loading in clarifier simulations, demonstrating that clarifier design is no longer empirical but increasingly data- and model-driven. These developments underscore the evolving role of secondary clarifiers not just as passive sedimentation basins, but as active regulators of biological treatment efficiency.

3.6 Membrane Bioreactor System

MBRs technology integrates biological wastewater treatment with membrane filtration. In this process, the degradation of organic matter takes place within a bioreactor tank, while the separation of treated effluent from microorganisms is achieved through a membrane module. From the early beginning of the twenty-first century, MBRs have gained considerable attention for their ability to consistently produce high-quality effluent and are now recognized as a mature and reliable wastewater treatment solution [40]. Given the limitations of conventional treatment methods, innovative alternatives like MBRs are being increasingly adopted. Furthermore, the modular and compact design of MBR systems has facilitated their use in decentralized and urban wastewater treatment applications, enhancing their global acceptance and versatility [41].

3.6.1 Principle of MBR

The fundamental principle of MBR technology lies in the integration of biological breakdown of organic matter in a bioreactor using microorganisms with membrane based solid–liquid separation within a single system. Unlike conventional activated sludge processes where sedimentation in a secondary clarifier is used to separate biomass from treated water, MBRs utilize membrane modules to directly retain suspended solids and microorganisms in the bioreactor. This physical barrier ensures a complete and consistent separation, resulting in an effluent of significantly higher quality. Biodegradation of organic matter occurs in the aeration tank, where activated biomass is maintained under aerobic conditions to break down pollutants. The membranes act as a filtration unit that allows only treated water to pass through while retaining MLSS and microbial flocs. This setup allows MBRs to operate at much higher biomass concentrations than conventional systems [42] (8000–12,000 mg/L), improving treatment efficiency and reducing the reactor volume needed. The elimination of the secondary clarifier and the enhanced SRT contribute to stable performance, better nutrient removal, and lower sludge production. The result is a compact, modular, and highly effective wastewater treatment solution capable of meeting stringent discharge standards. This enables the MBR to produce high-quality effluent suitable for reuse without further tertiary treatment [42].

3.6.2 MBR Design and Configuration

3.6.2.1 Membrane Materials

The membrane module is a key component of the MBRs and its material greatly influence reactor performance and membrane fouling. Commonly used membrane materials include polymers like PVDF, PES, PE, PP, PSF, CPE, and PTFE, as well as inorganic materials like ceramics [43–45]. Among these, PVDF membranes are the most widely used (~57% market share) due to their strong mechanical and chemical resistance [46]. PVDF also shows better control over irreversible fouling than PE, and its performance can be enhanced through modification with PFPE. Inorganic membranes, including stainless steel and ceramic types, offer high COD removal, better filtration, and longer service life. However, challenges like struvite precipitation and high production costs limit their application. With ongoing advances and cost reductions in manufacturing, the use of ceramic membranes is expected to grow significantly in the near future [46].

3.6.2.2 MBR Design

MBRs can be used as either aerobic or anaerobic bioreactors. Aerobic MBRs require aeration for microbial activity and fouling control, while anaerobic MBRs (AnMBRs) operate without oxygen and produce biogas as a valuable byproduct. MBRs are made up of two parts, (i) conventional anaerobic biological reactor and (ii) membrane modules. Based on the location of these two components, MBRs are divided into three categories: Fig. 3.4 (a) side-stream AnMBR, (b) internal/submerged AnMBR, and (c) external submerged AnMBR.

In the side-stream configuration, a recirculation pump generates transmembrane pressure and high shear forces, helping to reduce membrane fouling. However, this setup consumes more energy and may cause sludge floc breakage, releasing EPS/SMP that worsen fouling. An advantage is that membranes can be cleaned in-situ without stopping the system [47, 48]. In contrast, the internal submerged configuration places membranes directly in the bioreactor, using biogas recirculation instead of pumps for shear [49]. This reduces energy use but can increase dissolved methane in the effluent and makes chemical cleaning difficult, as membranes must be removed from the reactor. The external submerged configuration places membranes in a separate tank, allowing better shear control in a smaller volume. While this helps manage fouling, it still faces operational challenges and design complexity.

3.6.3 Performance and Benefits

MBRs offer significant advantages over conventional activated sludge systems, particularly in terms of high effluent quality, high biomass concentration, less rector volume, less footprint, less sludge production [50]. The effluent quality achieved by MBRs in terms of turbidity and suspended solids often falling below detection limits. The membrane's fine pore structure ensures high pathogen removal, including viruses and protozoa, resulting

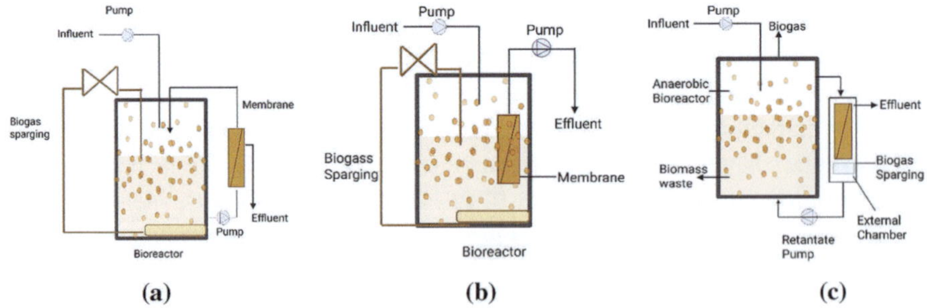

Fig. 3.4 Different configurations of AnMBRs **a** side-stream AnMBRs, **b** internal/submerged AnMBRs, and **c** external submerged AnMBRs

in a disinfection-grade effluent suitable for agricultural irrigation, industrial reuse, or even potable reuse after advanced treatment stages.

Several studies have demonstrated the strong performance of MBRs systems in wastewater treatment. Abegglen et al. [51] tested an on-site MBR in the basement of a house in Switzerland using flat sheet membranes. During the optimized phase, over 70% phosphorus and 90% nitrogen removal were achieved. In one more study, Abegglen and Siegrist [52] evaluated a small-scale MBR treating domestic wastewater over six months, achieving 80% total nitrogen (TN), 95% COD, and 90% TOC removal, indicating permeate quality comparable to conventional plants. In Canada, a 200 L pilot-scale MBR was tested in the Opaskwayak Cree Nation under cold conditions and still achieved high nitrification. Reductions in BOD (from 106 to 6 mg/L), TP (from 4 to 1.9 mg/L), TSS (from 79 to < 5 mg/L), and TKN (from 25 to 1.3 mg/L) were observed with a 10 h HRT and 20-day SRT [53]. Similarly, a full-scale MBR in Englewood, Colorado consistently met treated wastewater criteria, achieving CBOD < 5 mg/L, TSS < 5 mg/L, TP 0.2 mg/L, and TN < 10 mg/L after the system developed sufficient nitrifying biomass [54].

These findings highlight MBRs' effectiveness in nutrient and solids removal across varying scales and conditions, making them a promising option for decentralized wastewater treatment, though operational challenges are there for MBRs and are discussed in next section.

3.6.4 Challenges and Limitations

MBRs offer many advantages but face several challenges, fouling being most critical. Along with this, other significant challenges like high capital and operating costs, energy consumption, and the need for improved nutrient removal must also be addressed for MBRs to achieve their full potential [55].

3.6.4.1 Membrane Fouling

Fouling is a major concern in any membrane processes, as it significantly reduces performance, increases maintenance, and shortens membrane lifespan. Membrane fouling can be broadly categorized into three types based on the nature of the foulants [56]: inorganic, organic, and biological (bio-foulants). Inorganic fouling results from the buildup of minerals and fine particles such as silica, aluminum silicates, and iron-based compounds. Inorganic fouling is common in applications like seawater desalination and water softening. Organic foulants mainly consist of oils, large biological molecules like proteins, and natural organic matter such as humic acids and alginic acids. Biofouling, on the other hand, is caused by living organisms, including bacteria like E. coli and Staphylococcus aureus, as well as algae. Biofouling is the commonly observed in MBRs and is main cause of fouling in them [57]. Although inorganic fouling is widespread, it is typically managed through proper pre-treatment of the feed water or by optimizing operational conditions. Therefore, the development of antifouling membranes is primarily aimed at reducing organic and biological fouling, which are more challenging to prevent and clean once they occur.

Classification of fouling can be carried out based on different criteria like; flux recovery, location of fouling and solid deposition pattern. The simplest and most common criteria of flux recovery classifies fouling in three types reversible, irreversible and irrecoverable fouling [42]. Reversible fouling is the type of fouling where the water flux can be easily restored using basic cleaning methods like air scouring, pressure relaxation, or backwashing. While in irrecoverable fouling, these physical cleaning methods are not enough to restore the flux. As the fouling involves deeper layers like gel or materials stuck inside the membrane pores, chemical cleaning is often needed. Irreversible fouling is the most severe, where even chemical cleaning cannot recover the lost flux.

There is no unified statement about the mechanisms of membrane fouling, but from the analysis of the causes of membrane fouling, in MBRs systems, membrane fouling arises through several key mechanisms [58]:

(i) Adsorption of solutes or colloidal particles onto or within the membrane pores,
(ii) Deposition of sludge flocs on the membrane surface,
(iii) Formation of a cake layer due to the accumulation of retained solids,
(iv) Foulant detachment driven primarily by shear forces acting on the membrane, and
(v) Changes in foulant composition over time, including shifts in microbial communities and biopolymer content within the cake layer during prolonged operation.

Due to the complex composition of activated sludge, fouling behavior in MBRs is notably more complicated than in most of the other membrane applications. Membrane fouling in MBRs is mainly caused by sludge cake buildup and pore clogging which leads to reduced permeate flux or increased transmembrane pressure (TMP), depending on the operating mode. The fouling process in MBRs typically progresses through three distinct stages:

Stage 1: A rapid, short-term increase in TMP,

Stage 2: A prolonged period of gradual TMP rise,

Stage 3: A sudden and steep rise in rate of TMP change.

The primary foulants in MBRs are microbial species, predominantly extracellular polymeric substances (EPS) and soluble microbial products (SMP). EPS form a dense gel-like layer on the membrane surface, contributing to increased filtration resistance and promoting the accumulation of biofilm. SMP, being more soluble, penetrate the membrane structure and cause irreversible fouling by blocking the pores from within. Both EPS and SMP are by-products of microbial metabolism and decay, and their production is influenced by operational parameters like sludge retention time (SRT), hydraulic retention time (HRT), nutrient availability, and shear conditions membrane fouling and the life of membranes. The rate and intensity of fouling are influenced by several factors including MLSS concentration, aeration intensity, membrane module orientation, and membrane surface characteristics. Hydrodynamic conditions (e.g., cross-flow velocity, aeration scour) are also critical in controlling the thickness and compressibility of the cake layer.

3.6.4.2 Maintenance and Cleaning

Regular maintenance of MBRs and periodic membrane replacement is essential to managing the adverse effects of membrane fouling thereby significantly influencing membrane performance. Membrane cleaning is generally categorized into physical cleaning, chemical cleaning and biological cleaning [59]. Physical cleaning methods includes membrane relaxation (pausing filtration), backwashing using permeate or air at higher pressures. On the other hand, chemical cleaning includes use of common reagents like acids, bases, oxidants as well as some complex reagents like metal chelating chemicals, surfactants and formulated detergents. Chemical cleaning in MBRs systems is typically performed every 1 to 6 weeks and takes about 1 to 3 h to complete. Commonly used chemical concentrations for this cleaning include 300–2000 mg/L of sodium hypochlorite (NaClO) and 0.2–1.5% (by weight) of citric acid [60]. The combination of physical and chemical cleaning referred as physico-chemical cleaning is also a powerful pathway to improve membrane cleaning efficiency Two types of physico-chemical cleaning methods viz., chemically enhanced backwash (CEB) and ultrasound-aided chemical cleaning can also be employed for membrane cleaning. CEB needs to be carried out on a daily basis with 100–500 mg/L NaClO.

3.6.4.3 Energy and Cost Considerations

One of the key issues of MBRs is the high energy consumption, particularly for the aeration used in both biological processes and membrane scouring [55]. MBRs consumes significantly higher energy per cubic meter of treated wastewater than conventional activated sludge systems. However, newer plants have shown some improvements in energy efficiency through better system design and control strategies. There is still need to develop more energy-efficient configurations, such as advanced aeration strategies and optimized membrane-bioreactor integration.

Another important challenge associated with MBRs is the high capital and operating costs. The capital cost, which includes expenses related to construction, membrane modules, equipment, and infrastructure, remains considerably higher than that of conventional activated sludge systems. Although the cost becomes comparable when tertiary treatment is added, MBRs still require significant investment. Furthermore, operating costs are also elevated due to substantial energy demands, chemical usage for cleaning and nutrient removal, and labor expenses. Reducing these costs without compromising performance is critical to making MBRs more competitive.

3.6.5 Recent Advances in MBR

As already discussed, fouling is the primary challenge in MBRs, making the overall efficiency of these systems highly reliant on effective fouling control. Research on antifouling strategies remains a crucial area of focus for improving the efficiency and reliability of MBRs. Various strategies have been proposed for fouling management, such as the development of novel materials with enhanced antifouling properties, module designs that facilitate easier cleaning, and the formulation of control strategies to prevent fouling. Enhancing MBR systems through integration with advanced technologies is also receiving growing attention, as it can considerably improve overall performance in industrial wastewater treatment.

3.6.5.1 Novelties in Membrane Materials and Technology

Recent advances in materials science has led to application of advanced materials, like carbon nanotubes, graphene, and nanoparticles to reduce membrane fouling, thereby enhancing MBR efficacy [61]. Chemical grafting, surface coating, and blending with nanomaterials offer effective strategies for developing antifouling membranes. Mixed matrix membranes (MMMs), created by embedding nanoparticles into polymers via self-assembly, are gaining attention for their hydrophilicity, antimicrobial properties, and photocatalytic self-cleaning. Membranes incorporating multidimensional carbon-based materials such as nanospheres, nanofibers, nanotubes, and nanosheets offers enhanced permeability. Compared to conventional nanoparticles, these nanomaterials possess superior hydrophilicity and antimicrobial properties. Blending membranes with metal or inorganic oxides like TiO_2, Fe_3O_4, Cu_2O, and SiO_2 has gained significant interest due to their distinct functionalities and stable performance. Silver nanoparticles (Ag NPs), known for their strong antimicrobial activity, are also commonly integrated into polymeric membranes to improve anti-biofouling performance [62]. Algal-based MBR (AMBRs) are novel types of wastewater treatment reactors that combine microalgae cultivation with membrane filtration MBRs have shown the strong potential for removing hazardous contaminants from wastewater, offering high efficiency, low energy use, and by-product recovery. However, there are certain challenges of MBRs like membrane fouling, microbial inhibition, nutrient and oxygen limitations which must be addressed for optimizing their performance [63].

One new and effective approach to reduce fouling is modifying the MBRs configuration. Dynamic membrane (DM) technology has emerged as a promising addition to MBRs systems, offering benefits like strong antifouling ability, low-cost support materials, and easy cleaning [64]. Instead of relying on traditional aeration for fouling control, DM approach use movable membrane modules to create stronger shear forces that more effectively reduce fouling and enhance MBRs performance. The DM configurations includes Vibrating membrane, reciprocating membrane, rotating membrane and implanted ends-free MBRs (IEF-MBRs) [64]. Several advanced methods like quorum quenching, electrochemistry, and mechanically imposed membrane shear are used to disrupt microbial growth on membranes thereby controlling the membrane biofouling [57]. Notable advancements include the use of Artificial Intelligence (AI)-based models, such as neural networks and random forests, to predict membrane fouling and optimize operational parameters in MBR systems [65].

3.6.5.2 Hybrid and Integrated MBRs

Enhancing MBRs performance through integration with advanced technologies is gaining attention for its ability to improve industrial wastewater treatment and reduce membrane fouling [66]. Combining MBRs with microbial fuel cells (MFCs) presents a promising hybrid approach due to low energy consumption and enhanced treatment efficiency. In this setup, conductive membranes function both as filtration media and MFC cathodes, enabling sludge reduction, fouling control, and energy recovery. The electrostatic repulsion between foulants and cathode membrane surface resulted in membrane fouling reduction [67]. Moreover, MBRs integration MFC modifies sludge characteristics, improving filterability by increasing floc size and decreasing soluble microbial products [68]. Catalytic MFC-MBR systems using nanomaterials such as $CoFe_2O_4$-rGO have demonstrated superior oxygen reduction activity, conductivity and permeability under visible light [69].

Forward osmosis (FO) integrated with MBRs (OMBRs) enhances water quality and energy efficiency. FO draws treated water from mixed liquor into a concentrated draw solution, which is later regenerated. Porous FO membranes exhibit higher water flux and rejection of organic pollutants compared to dense ones [70]. However, reverse salt diffusion leads to salinity build-up, negatively impacting microbial activity and flux. Solutions include the development of high-rejection, fouling-resistant FO membranes and alternative draw solutes with low internal concentration polarization [71]. AOPs like ozonation, photocatalysis, and Fenton reactions are widely used for treating refractory pollutants. Ozonation enhances MBRs permeability performance by increasing particle size and membrane hydrophobicity [72]. A theoretical economic analysis for pesticides treatment on large scale using combined Photo-Fenton with MBR was a very effective approach [73].

3.6.5.3 Real-World Applications

MBRs have been widely adopted globally for municipal and industrial wastewater treatment due to their compact design, high effluent quality, and adaptability. Installations span UK (Porlock, Swanage), USA (Dundee, Running Springs), Spain (Sabadell), South Africa (Illovo Sugar), and Oman (Al Ansab), each tailored to specific site constraints and treatment goals [41]. Several global companies such as Koch, Siemens (Evoqua/Veolia), Kubota, Veolia, Sumitomo, Motimo, WEHRLE, Norit X-Flow, Dynatec, Memstar, Weise Water Systems, Triqua, and HERA AMASA have installed or supplied MBRs systems worldwide, covering municipal and industrial applications using technologies like submerged flat-sheet, hollow fiber, multitube, and catalytic membranes. Pre-treatment usually includes fine screening, grit removal, and flow equalization. Plants operate under varying hydraulic and organic loads, with MLSS levels often above 10 g/L, and fluxes ranging from 15–30 LMH. Energy efficiency and maintenance protocols such as air scouring, membrane relaxation, and periodic chemical cleaning ensure long-term reliability.

A survey conducted in 2022 reported 290 MBRs municipal or industrial WWTPs are in operation in Italy [74]. China have rapidly expanded its large-scale MBRs applications since the first major installation in 2006, reaching a cumulative capacity of 1 million m^3/d by 2010 and 10 million m^3/d by 2017 [75]. As of 2021, the largest MBR plant in operation was the Beihu Wastewater Treatment Plant in Hubei, China, commissioned in 2019, with a design average inflow of 9.3 m^3/s which significantly exceeds the future Henriksdal WWTP's design capacity of 6.1 m^3/s [75]. The challenges for the MBRs process at Henriksdal WWTP include managing seasonal variations in temperature and flow that impact membrane efficiency and nitrogen removal, achieving strict phosphorus limits through precipitation without compromising membranes, and minimizing overall resource consumption. A recent study suggested that the alternative membrane cleaning strategies at the Henriksdal plant can reduce the chemical use by up to 75% without sacrificing treatment performance [76].

In conclusion, MBRs represent a versatile and high-performance option for secondary wastewater treatment. While there are certain challenges like membrane fouling and energy cost, continuous research and technological advancements are making MBRs more feasible for large-scale applications. Future research progress focusses on the development of cost-effective membranes, novel design, integration with renewable energy sources, and implementation of AI based control systems to ensure sustainable operation. The resulting improvements will help to ensure that MBRs run smoothly and in a more environmentally friendly way.

3.7 Effluent Disinfection and Dechlorination

Many of the wastewater treatment plant designed for treating municipal wastewater or wastewater having no toxic pollutant discharge treated water after secondary treatment itself. Although, secondary treatment significantly reduces organic load and suspended solids, it is not sufficient in reducing the microbial hazard of treated effluent. Such effluent is dangerous when the water is intended for reuse or discharge into sensitive environments. The secondary effluent contains mainly four types of human enteric organisms [77] (1) bacteria, (2) viruses, (3) protozoa, and (3) helminth ova that persists even after biological treatment. These microorganisms can several diseases some of which are life threatening [78]. Hence it is important to disinfect such pathogenic microorganism containing secondary effluent to eliminate them.

The disinfection process primarily works through the disruption of cellular membranes, denaturation of proteins, and damage to the genetic material of microorganisms. This either inactivates or destroys pathogens, rendering them incapable of replication or infection. The disinfection of secondary effluent can be achieved either by chemical agents or by non-ionizing radiation or by both. Common chemical disinfectants include chlorine, hypochlorite's, ozone, bromine, iodine, and hydrogen peroxide, while no iodizing radiation includes UV irradiation [13]. The choice of disinfection method is influenced by factors such as effluent quality, regulatory standards, and intended reuse of water.

3.7.1 Chlorination

Chlorination, either achieved by chlorine gas or by chlorine compounds is the most widely adopted method for wastewater disinfection. Chlorine used for disinfection can be a gas or a pressurized liquid. Chlorine gas is yellow in color and is approximately 2.48 times as heavy as air. While, liquid chlorine is yellowish-brown color and about 1.44 times as heavy as water. When chlorine in the form of Cl_2 gas is added to water, two reactions take place: hydrolysis and ionization [79].

$$\text{Hydrolysis}: Cl_2 + H_2O \rightarrow HOCl + H^+ + Cl^-$$

The equilibrium constant K_H for this reaction is

$$K_H = \frac{[HOCl][H^+][Cl^-]}{[Cl_2]} = 3.94 \times 10^{-4}\left(\frac{mol^2}{l^2}\right) at\, 25°C \quad (3.3)$$

Ionization: $HOCl \leftrightarrow H^+ + OCl^-$

The ionization constant K_H for this reaction is

$$K_H = \frac{[HOCl][H^+][Cl^-]}{[Cl_2]} = 2.9 \times 10^{-8}\left(\frac{mol}{l}\right) at\, 25°C \quad (3.4)$$

The total quantity of HOCl and OCl^- present in water is called the free chlorine. The relative distribution of these two species is very important because HOCl is far more effective as a disinfectant than OCl^-. HOCl is uncharged, allowing it to penetrate microbial cell walls more easily. Once inside the cell, chlorine oxidizes and denatures proteins, damages nucleic acids to prevent replication, and disrupts membranes, leading to cell death. Most of the chlorination results in free chlorine remained in the water post-treatment, preventing microbial regrowth during storage or distribution as well.

The rate of change in the concentration (number) of organisms with time using Chick and Watson law is given as:

$$\frac{dN_t}{dt} = -\Lambda_{CW} C^n N_t \tag{3.5}$$

where Λ_{CW} is the Chick-Watson coefficient of specific lethality,

C is the residual concentration of disinfectant,

n is the empirical constant related to dilution.

If n is equal to one, the above equation can be integrated and written as:

$$log \frac{N_t}{N_0} = -\Lambda_{CW\,base10}(CT) \tag{3.6}$$

where T is the contact time in the reactor.

CT value (also called as dose) which is the product of the (C) concentration of a disinfectant and (T) the contact time with the water being disinfected is an important part of calculating disinfectant dosage for the chlorination of drinking water. Chlorine's effectiveness depends on several factors like pH, CT value and presence of organic/inorganic matter. It increases with decrease in pH and adversely gets affected due to present of organic/inorganic matter which reduces free available chlorine.

Chlorination has several advantages like higher disinfection efficacy, low cost, simple operation and are well established methods. Chlorine compounds can effectively inactivate a wide range of pathogens at relatively low doses within short contact times. Also, chlorination offers continued microbial control throughout storage or distribution. However, it also has few disadvantages mainly the formation of disinfection by-products (DBPs) such as trihalomethanes (THMs) and haloacetic acids (HAAs) after series of complex reactions [80]. Many of these DBPs, Chloroform, for example, are potentially carcinogenic. Additionally, chlorine is highly toxic and corrosive, posing safety hazards during transport, storage, and handling. Factors such as pH, temperature, chlorine dose, contact time, and the presence of ammonia or organic matter significantly influence its effectiveness. It also requires specialized containment infrastructure due to its corrosive and volatile nature. The risk of accidental release has led many facilities to shift away from gaseous chlorine.

Instead of chlorine gas, HOCl can be formed in water by addition of hypochlorite. Both sodium hypochlorite (NaOCl) and calcium hypochlorite ($Ca(OCl)_2$) gives HOCl by hydrolysis as follows:

$$NaOCl + H_2O \rightarrow HOCl + NaOH$$

$$Ca(OCl)_2 + H_2O \rightarrow 2HOCl + Ca(OH)_2$$

NaOCl, commonly known as liquid bleach, is available as an aqueous solution and is safer to handle compared to chlorine gas. Solution of NaOCl usually contains about 12.5% to 17% chlorine at the time it is manufactured. The NaOCl solution decomposes more readily at high concentrations as it is affected by exposure to light and heat. It must, therefore, be stored in a cool location in a corrosion resistant tank to avoid loss of chlorine strength. Another disadvantage of sodium hypochlorite is the chemical cost often costing up to twice as much as chlorine gas. Moreover, it can produce chlorine fumes, and may cause gas binding in feed lines, requiring careful design considerations.

$Ca(OCl)_2$ is commercially available in either dry or a wet form. It is available in granular, tablet, or pellet forms and is readily soluble in water, varying from about 21.5 g/ 100 mL at 0 °C to 23.4 g/100 mL at 40 °C. It is generally stable under cool and dry conditions and is fairly easy to handle. However, due to its strong oxidizing nature, it must be stored in a cool, dry place, separate from other chemicals. It is more expensive than liquid chlorine and needs to be dissolved before application as it may clog equipment because of crystallization. As a result, its use is generally restricted to small-scale or decentralized treatment systems.

3.7.2 Ozonation

Ozonation is an effective chemical disinfection method that employs ozone (O_3) to rapidly inactivate a wide range of microorganisms, including bacteria, viruses, and protozoa. Ozone is blue-colored gas with a distinct odor that can be detected by humans even at low concentrations (0.01 to 0.05 ppm). Historically, in the U.S., ozonation was evaluated for wastewater treatment starting in the 1970s [81]. Early systems faced operational difficulties, and the technology was often viewed as less favorable compared to UV or chlorination. However, recent advancements in ozone generation and dissolution technology have made ozone a more cost-effective option for disinfecting treated wastewater. Further, ozone's ability to reduce or eliminate trace contaminants has increase interest in ozonation for disinfection.

Ozone is very strong oxidizing agent which penetrates the microbial cell membrane. It oxidizes essential biological components such as lipids, proteins, enzymes, and DNA, ultimately leading to cell lysis and death. One of the key advantages of ozonation is that it does not produce harmful by-product similar to chlorination and leaves no residual disinfectant in the treated water. Another advantage of Ozone disinfection is that it raises effluent dissolved oxygen to near saturation, often eliminating the need for additional reaeration to meet water quality standards. Ozonation does not raise dissolved solids, and its disinfection efficiency remains unaffected by the presence of ammonium ions. Ozone is unstable and it decomposes within minutes in both air and water. On decomposition it forms OH• radicals which are the strongest oxidants in water. For different applications of ozone, the both Ozone and OH• are of differing importance, while disinfection occurs specifically through ozone, oxidation may occur through both [82].

Due to its high instability, it is usually generated at onsite commonly through electrical discharge, which is the standard method used in water and wastewater treatment applications. Other methods of ozone generation include electrolysis, photochemical, and radiochemical reactions. A major operational challenge in wastewater ozonation is ensuring efficient mass transfer of ozone from the gas phase into the water, as this step critically affects microbial inactivation. The fast reaction kinetics between ozone and fecal coliforms suggest that contact time is less significant than the amount of ozone actually transferred [81]. Modern ozonation systems emphasize short contact times with improved ozone transfer efficiency to optimize performance and cost-effectiveness. Despite its effectiveness, ozonation has few limitations. It requires high energy input and involves expensive installation and maintenance.

3.7.3 Ultraviolet Disinfection

Ultraviolet (UV) radiation induced disinfection is a method which unlike traditional disinfection does not require the addition of chemicals. UV disinfection works within the germicidal range of 200 – 300 nm wavelengths and it is particularly effective against chlorine-resistant pathogens like *Giardia* and *Cryptosporidium* [12]. However, viruses particularly adenoviruses and bacterial spores are the organisms which are most resistant to UV rays [83]. Low-pressure (LP-UV) lamps emit light at 254 nm which is absorption peak of nucleic acids. These LP-UV irradiation damages the DNA and RNA of cell primarily through the formation of pyrimidine dimers [84]. If unrepaired, this damage halts replication and transcription, inactivating microorganisms. Medium-pressure (MP-UV) lamps emit broader wavelengths that can also damage proteins and viral structures. The effectiveness of UV disinfection depends on wastewater chemistry, particle presence, microorganism type, and system design. The effectiveness of UV disinfection is based on the UV dose to which the microorganisms are exposed. The UV dose, D, which is analogous to dose term CT is given by;

$$D = I_{avg} \times t \tag{3.7}$$

where

D = UV dose, (mJ/cm^2).

I_{avg} = average UV intensity, mW/cm^2

t = exposure time.

This UV dose can be varied either by changing intensity or by time of exposure. UV disinfection operates with much shorter hydraulic residence time of 5–30 s at UV doses of 30–60 mJ/cm^2 as against chlorination, which requires a contact time of 20–60 min at a dose of 0.5 mg/L to have similar microbial inactivation [85]. UV disinfection produces no harmful by-products and also avoids the possible generation of disinfection resistance in bacteria [86].

However, UV disinfection comes with few disadvantages as well. UV disinfection does not provide a lasting residual effect, and certain pathogenic microorganisms have been shown to regain activity upon exposure to light [87]. This phenomenon is also called photoreactivation. Effectiveness of UV disinfection is compromised by turbidity or suspended solids that can shield pathogens. It also requires regular lamp maintenance and an uninterrupted power supply, making it more operationally intensive in certain settings. Also, traditional UV lamps contain mercury in a bulky and fragile glass vacuum tubes for irradiating UV light. This makes handling of UV lamp and disposal of remains an environmental burden. However, UV light emitting diodes (UV LED) lamps offers alternative to mercury-based lamp with advantages of tremendous design flexibility and no toxic wastes. Over the past decade there have been rapid advancement in UV-LED technology that have transitioned large-scale implementation from concept to reality. With ongoing improvements in cost, power output, and efficiency, LED-based disinfection is increasingly becoming viable for water and wastewater treatment overcoming few of the shortcomings of conventional UV disinfection [88].

Apart from using these methods individually, combinations of Chlorination/O_3/UV with other techniques can be employed. These combinations not only achieve microbial disinfection but also contribute to the degradation of dissolved organic compounds. These combined methods are discussed in the next chapter. The use of any disinfection systems for particular plant must weigh in their advantages and disadvantages. Site-specific factors such as water quality parameters, hydraulic conditions, and maintenance infrastructure to need to be considered to ensure consistent microbial deactivation. Nevertheless, disinfection remains an essential process in safeguarding public health and environmental quality.

3.7.4 De-Chlorination

Although various disinfection techniques are available, chlorination is still most commonly used methods for the destruction of pathogenic and other harmful organisms. However, certain organic constituents in wastewater react with the chlorine forming toxic compounds that can have long-term adverse effects on environment or human. Thus to reduce the environmental impact of toxic chlorine residuals, treated effluent must be dechlorinated. This dechlorination is carried out using reducing agents like sulfur dioxide or sodium bisulfite, or through adsorption with activated carbon. When effluent toxicity limits apply or dechlorination is used as polishing step to remove ammonia, sulfur dioxide is commonly used for dechlorination. Sulfur dioxide forms sulfurous acid in water, a strong reducing agent. It dissociates into bisulfite ions that reduce free and combined chlorine to chloride and sulfate.

$$SO_2 + H_2O \rightarrow H_2SO_3$$

$$H_2SO_3 + HOCl \rightarrow HCl + H_2SO_4$$

Sulfur dioxide also removes monochloramine, dichloramine, and nitrogen trichloride:

$$NH_2Cl + H_2SO_3 + H_2O \rightarrow NH_4Cl + H_2SO_4$$
$$NHCl_2 + 2H_2SO_3 + 2H_2O \rightarrow NH_4Cl + 2H_2SO_4 + HCl$$
$$NCl_3 + 3H_2SO_3 + 3H_2O \rightarrow NH_4Cl + 3H_2SO_4 + 2HCl$$

Sulfur dioxide dechlorination is generally reliable if combined chlorine residuals are monitored accurately. However, overdosing of sulfur dioxide should be avoided to prevent chemical waste and increased oxygen demand, as excess sulfur dioxide reacts slowly with dissolved oxygen.

$$HSO_3^- + 0.5O_2 \rightarrow SO_4^- + H^+$$

This will reduce the dissolved oxygen leading to increase in measured BOD and COD. Therefore, careful control of sulfur dioxide dosing is essential to ensure effective dechlorination while minimizing residual oxygen demand and maintaining effluent quality. Proper monitoring ensures both regulatory compliance and environmental protection.

Secondary biological treatment is a cornerstone of modern wastewater management, harnessing the metabolic capabilities of microorganisms to degrade organic pollutants and remove nutrients effectively. The section explored fundamental principles such as microbial metabolism, biokinetics, and the dynamics of activated sludge microbiology, which are critical for understanding treatment performance. The activated sludge process, in its various configurations, remains the most widely implemented system due to its efficiency and adaptability, despite operational challenges like sludge management and process control. MBRs represent a significant advancement, offering superior effluent quality and compact design, though issues like membrane fouling and cost remain. Secondary clarifiers play a crucial role in solid–liquid separation, ensuring treated effluent meets discharge or reuse standards. Finally, disinfection methods such as chlorination, UV, and ozonation provide the final barrier against pathogens. Overall, this stage not only ensures compliance with environmental regulations but also enables safe effluent reuse, underscoring its essential role in sustainable wastewater treatment systems.

References

1. Mathanmohun, M.; Elango, B.; Okram, G. S.; Shah, M. P. Potentials of Microbes in Wastewater Treatment and Management. In *Recalcitrant Pollutants Removal from Wastewater*; CRC Press, 2024; pp 153–179.
2. Ghangrekar, M. M. Aerobic Wastewater Treatment Systems. In *Wastewater to Water*; Springer Nature Singapore: Singapore, 2022; pp 395–474. https://doi.org/10.1007/978-981-19-4048-4_10.
3. Ghangrekar, M. M. Anaerobic Process for Wastewater Treatment. In *Wastewater to Water*; Springer Nature Singapore: Singapore, 2022; pp 507–553. https://doi.org/10.1007/978-981-19-4048-4_12.

4. Madakka, M.; Jayaraju, N.; Rajesh, N.; Chandra, M. R. G. S. Development in the Treatment of Municipal and Industrial Wastewater by Microorganism. In *Recent developments in applied microbiology and biochemistry*; Elsevier, 2019; pp 263–273.
5. Kumar, A.; Chalotra, T.; Pathak, A. K.; Raina, N. Role of Microbes in Wastewater Treatment and Energy Generation Potentials. In *Microbial Biotechnology*; Wiley, 2022; pp 265–311. https://doi.org/10.1002/9781119834489.ch15.
6. Mbamalu Ezeh, E.; Chinedu Agu, P.; Aworabhi, E. Biological Treatment Techniques for Sewage: Aerobic and Anaerobic Processes. In *Sewage - Management and Treatment Techniques*; IntechOpen, 2024. https://doi.org/10.5772/intechopen.1006097.
7. Rahimi, S.; Zinatizadeh, A. A.; Mohammadi, P.; Zinadini, S.; Asadi, A. Performance of an Activated Sludge Followed by Membrane Process (AS-MP) Treating Simulated Industrial Wastewaters: Effects of Operating Factors and Feed Characteristics. *Appl. Water Sci.* **2020**, *10* (8), 203. https://doi.org/10.1007/s13201-020-01285-1.
8. Vítězová, M.; Kohoutová, A.; Vítěz, T.; Hanišáková, N.; Kushkevych, I. Methanogenic Microorganisms in Industrial Wastewater Anaerobic Treatment. *Processes* **2020**, *8* (12), 1546. https://doi.org/10.3390/pr8121546.
9. Singh, R.; Pareek, N.; Kumar, R.; Vivekanand, V. Anaerobic Biodigesters for the Treatment of High-Strength Wastewater; 2022; pp 75–94. https://doi.org/10.1007/978-981-19-4921-0_5.
10. Zamri, M. F. M. A.; Hasmady, S.; Akhiar, A.; Ideris, F.; Shamsuddin, A. H.; Mofijur, M.; Fattah, I. M. R.; Mahlia, T. M. I. A Comprehensive Review on Anaerobic Digestion of Organic Fraction of Municipal Solid Waste. *Renew. Sustain. Energy Rev.* **2021**, *137*, 110637. https://doi.org/10.1016/j.rser.2020.110637.
11. Monod, J. THE GROWTH OF BACTERIAL CULTURES. *Annu. Rev. Microbiol.* **1949**, *3* (1), 371–394. https://doi.org/10.1146/annurev.mi.03.100149.002103.
12. Metcalf & Eddy; Abu-Orf, M.; Bowden, G.; Burton, F. L.; Pfrang, W.; Stensel, H. D.; Tchobanoglous, G.; Tsuchihashi, R.; AECOM (Firm). *Wastewater Engineering: Treatment and Resource Recovery*; McGraw Hill Education, 2014.
13. Mason, I. Mathematical Modelling of the Composting Process: A Review. *Waste Manag.* **2006**, *26* (1), 3–21.
14. Henze, M.; Gujer, W.; Mino, T.; Van Loosedrecht, M. Activated Sludge Models ASM1, ASM2, ASM2d and ASM3. *Water Intell. Online* **2015**, *5* (0), 9781780402369–9781780402369. https://doi.org/10.2166/9781780402369.
15. Chauhan, A. S.; Kumar, A.; Parmar, K.; Kumar, V. Microbial Ecology of Wastewater Treatment Processes: Trends, Challenges, and Perspectives. In *Omics Insights in Environmental Bioremediation*; Springer Nature Singapore: Singapore, 2022; pp 301–315. https://doi.org/10.1007/978-981-19-4320-1_13.
16. Severo, I. A.; Siqueira, S. F.; Deprá, M. C.; Maroneze, M. M.; Zepka, L. Q.; Jacob-Lopes, E. Biodiesel Facilities: What Can We Address to Make Biorefineries Commercially Competitive? *Renew. Sustain. Energy Rev.* **2019**, *112*, 686–705. https://doi.org/10.1016/j.rser.2019.06.020.
17. Wu, L.; Ning, D.; Zhang, B.; Li, Y.; Zhang, P.; Shan, X.; Zhang, Q.; Brown, M. R.; Li, Z.; Van Nostrand, J. D.; Ling, F.; Xiao, N.; Zhang, Y.; Vierheilig, J.; Wells, G. F.; Yang, Y.; Deng, Y.; Tu, Q.; Wang, A.; Acevedo, D.; Agullo-Barcelo, M.; Alvarez, P. J. J.; Alvarez-Cohen, L.; Andersen, G. L.; de Araujo, J. C.; Boehnke, K. F.; Bond, P.; Bott, C. B.; Bovio, P.; Brewster, R. K.; Bux, F.; Cabezas, A.; Cabrol, L.; Chen, S.; Criddle, C. S.; Deng, Y.; Etchebehere, C.; Ford, A.; Frigon, D.; Sanabria, J.; Griffin, J. S.; Gu, A. Z.; Habagil, M.; Hale, L.; Hardeman, S. D.; Harmon, M.; Horn, H.; Hu, Z.; Jauffur, S.; Johnson, D. R.; Keller, J.; Keucken, A.; Kumari, S.; Leal, C. D.; Lebrun, L. A.; Lee, J.; Lee, M.; Lee, Z. M. P.; Li, Y.; Li, Z.; Li, M.; Li, X.; Ling, F.; Liu, Y.; Luthy, R. G.; Mendonça-Hagler, L. C.; de Menezes, F. G. R.; Meyers, A. J.; Mohebbi, A.; Nielsen, P. H.; Ning, D.; Oehmen, A.; Palmer, A.; Parameswaran, P.; Park, J.; Patsch, D.;

Reginatto, V.; de los Reyes, F. L.; Rittmann, B. E.; Noyola, A.; Rossetti, S.; Shan, X.; Sidhu, J.; Sloan, W. T.; Smith, K.; de Sousa, O. V.; Stahl, D. A.; Stephens, K.; Tian, R.; Tiedje, J. M.; Tooker, N. B.; Tu, Q.; Van Nostrand, J. D.; De los Cobos Vasconcelos, D.; Vierheilig, J.; Wagner, M.; Wakelin, S.; Wang, A.; Wang, B.; Weaver, J. E.; Wells, G. F.; West, S.; Wilmes, P.; Woo, S.-G.; Wu, L.; Wu, J.-H.; Wu, L.; Xi, C.; Xiao, N.; Xu, M.; Yan, T.; Yang, Y.; Yang, M.; Young, M.; Yue, H.; Zhang, B.; Zhang, P.; Zhang, Q.; Zhang, Y.; Zhang, T.; Zhang, Q.; Zhang, W.; Zhang, Y.; Zhou, H.; Zhou, J.; Wen, X.; Curtis, T. P.; He, Q.; He, Z.; Brown, M. R.; Zhang, T.; He, Z.; Keller, J.; Nielsen, P. H.; Alvarez, P. J. J.; Criddle, C. S.; Wagner, M.; Tiedje, J. M.; He, Q.; Curtis, T. P.; Stahl, D. A.; Alvarez-Cohen, L.; Rittmann, B. E.; Wen, X.; Zhou, J. Global Diversity and Biogeography of Bacterial Communities in Wastewater Treatment Plants. *Nat. Microbiol.* **2019**, *4* (7), 1183–1195. https://doi.org/10.1038/s41564-019-0426-5.

18. Xie, N.; Zhong, L.; Ouyang, L.; Xu, W.; Zeng, Q.; Wang, K.; Zaynab, M.; Chen, H.; Xu, F.; Li, S. Community Composition and Function of Bacteria in Activated Sludge of Municipal Wastewater Treatment Plants. *Water* **2021**, *13* (6), 852. https://doi.org/10.3390/w13060852.
19. Chen, Y.; Jiang, X.; Yang, M.; Wang, Z. Biotechnology Revival: In Situ Sludge Minimization in Wastewater. *Front. Microbiol.* **2025**, *16*. https://doi.org/10.3389/fmicb.2025.1603215.
20. Kirilova, M.; Yotinov, I.; Todorova, Y.; Dinova, N.; Lincheva, S.; Schneider, I.; Topalova, Y. Microbiome Structure of Activated Sludge after Adaptation to Landfill Leachate Treatment in a Lab-Scale Sequencing Batch Reactor. *Processes* **2024**, *12* (1), 159. https://doi.org/10.3390/pr12010159.
21. Tan, Z.; Chen, W.; Guo, Z.; Xu, X.; Xie, J.; Dai, J.; Lin, Y.; Sheng, B.; Preis, S.; Wei, C.; Zhu, S. Seasonal Dynamics of Bacterial Composition and Functions in Biological Treatment of Coking Wastewater. *Appl. Microbiol. Biotechnol.* **2024**, *108* (1), 490. https://doi.org/10.1007/s00253-024-13274-4.
22. Gerardi, M. H. *Nitrification and Denitrification in the Activated Sludge Process*, 1st ed.; Wiley, 2002. https://doi.org/10.1002/0471216682.
23. Ergas, S. J.; Aponte-Morales, V. Biological Nitrogen Removal. In *Comprehensive Water Quality and Purification*; Elsevier, 2014; pp 123–149. https://doi.org/10.1016/B978-0-12-382182-9.00047-5.
24. Mainstone, C. P.; Parr, W. Phosphorus in Rivers — Ecology and Management. *Sci. Total Environ.* **2002**, *282–283*, 25–47. https://doi.org/10.1016/S0048-9697(01)00937-8.
25. Barnard, J. L. Biological Nutrient Removal without the Addition of Chemicals. *Water Res.* **1975**, *9* (5–6), 485–490. https://doi.org/10.1016/0043-1354(75)90072-X.
26. Jenkins, D.; Wanner, J. *Activated Sludge-100 Years and Counting*; IWA publishing, 2014.
27. Fogler, H. S. *Elements of Chemical Reaction Engineering*; Pearson Educacion, 1999.
28. Mahvi, A. Sequencing Batch Reactor: A Promising Technology in Wastewater Treatment. **2008**.
29. Sam, T.; Le Roes-Hill, M.; Hoosain, N.; Welz, P. J. Strategies for Controlling Filamentous Bulking in Activated Sludge Wastewater Treatment Plants: The Old and the New. *Water* **2022**, *14* (20), 3223.
30. Bafghi, M. F.; Yousefi, N. Role of Nocardia in Activated Sludge. *Malays. J. Med. Sci. MJMS* **2016**, *23* (3), 86.
31. Holenda, B.; Domokos, E.; Rédey, Á.; Fazakas, J. Dissolved Oxygen Control of the Activated Sludge Wastewater Treatment Process Using Model Predictive Control. *Comput. Chem. Eng.* **2008**, *32* (6), 1270–1278. https://doi.org/10.1016/j.compchemeng.2007.06.008.
32. Ahansazan, B.; Afrashteh, H.; Ahansazan, N.; Ahansazan, Z. Activated Sludge Process Overview. *Int. J. Environ. Sci. Dev.* **2014**, *5* (1), 81.
33. Wang, L. K.; Wu, Z.; Shammas, N. K. Activated Sludge Processes. In *Biological treatment processes*; Springer, 2009; pp 207–281.

34. Guo, W.-Q.; Yang, S.-S.; Xiang, W.-S.; Wang, X.-J.; Ren, N.-Q. Minimization of Excess Sludge Production by In-Situ Activated Sludge Treatment Processes — A Comprehensive Review. *Biotechnol. Adv.* **2013**, *31* (8), 1386–1396. https://doi.org/10.1016/j.biotechadv.2013.06.003.
35. Ren, S. Assessing Wastewater Toxicity to Activated Sludge: Recent Research and Developments. *Environ. Int.* **2004**, *30* (8), 1151–1164. https://doi.org/10.1016/j.envint.2004.06.003.
36. Eikelboom, D. H. *Process Control of Activated Sludge Plants by Microscopic Investigation*; IWA publishing, 2000.
37. Gavrilescu, M.; Macoveanu, M. Attached-Growth Process Engineering in Wastewater Treatment. *Bioprocess Eng.* **2000**, *23* (1), 95–106.
38. Droste, R. L.; Gehr, R. L. *Theory and Practice of Water and Wastewater Treatment*; John Wiley & Sons, 2018.
39. Goula, A. M.; Kostoglou, M.; Karapantsios, T. D.; Zouboulis, A. I. A CFD Methodology for the Design of Sedimentation Tanks in Potable Water Treatment: Case Study: The Influence of a Feed Flow Control Baffle. *Chem. Eng. J.* **2008**, *140* (1–3), 110–121.
40. Marrot, B.; Barrios-Martinez, A.; Moulin, P.; Roche, N. Industrial Wastewater Treatment in a Membrane Bioreactor: A Review. *Environ. Prog.* **2004**, *23* (1), 59–68.
41. Judd, S. *The MBR Book: Principles and Applications of Membrane Bioreactors for Water and Wastewater Treatment*; Elsevier, 2010.
42. Al-Asheh, S.; Bagheri, M.; Aidan, A. Membrane Bioreactor for Wastewater Treatment: A Review. *Case Stud. Chem. Environ. Eng.* **2021**, *4*, 100109. https://doi.org/10.1016/j.cscee.2021.100109.
43. Skouteris, G.; Hermosilla, D.; López, P.; Negro, C.; Blanco, Á. Anaerobic Membrane Bioreactors for Wastewater Treatment: A Review. *Chem. Eng. J.* **2012**, *198–199*, 138–148. https://doi.org/10.1016/j.cej.2012.05.070.
44. Lin, H.; Peng, W.; Zhang, M.; Chen, J.; Hong, H.; Zhang, Y. A Review on Anaerobic Membrane Bioreactors: Applications, Membrane Fouling and Future Perspectives. *Desalination* **2013**, *314*, 169–188. https://doi.org/10.1016/j.desal.2013.01.019.
45. Liu, F.; Hashim, N. A.; Liu, Y.; Abed, M. R. M.; Li, K. Progress in the Production and Modification of PVDF Membranes. *J. Membr. Sci.* **2011**, *375* (1–2), 1–27. https://doi.org/10.1016/j.memsci.2011.03.014.
46. Raeesh, M.; Devi, T. T.; Hirom, K. Recent Developments on Application of Different Turbulence and Multiphase Models in Sedimentation Tank Modeling—a Review. *Water. Air. Soil Pollut.* **2023**, *234* (1), 5.
47. Chang, S. Anaerobic Membrane Bioreactors (AnMBR) for Wastewater Treatment. *Adv. Chem. Eng. Sci.* **2014**, *04* (01), 56–61. https://doi.org/10.4236/aces.2014.41008.
48. Dvořák, L.; Gómez, M.; Dolina, J.; Černín, A. Anaerobic Membrane Bioreactors—a Mini Review with Emphasis on Industrial Wastewater Treatment: Applications, Limitations and Perspectives. *Desalination Water Treat.* **2016**, *57* (41), 19062–19076. https://doi.org/10.1080/19443994.2015.1100879.
49. Maaz, M.; Yasin, M.; Aslam, M.; Kumar, G.; Atabani, A. E.; Idrees, M.; Anjum, F.; Jamil, F.; Ahmad, R.; Khan, A. L.; Lesage, G.; Heran, M.; Kim, J. Anaerobic Membrane Bioreactors for Wastewater Treatment: Novel Configurations, Fouling Control and Energy Considerations. *Bioresour. Technol.* **2019**, *283*, 358–372. https://doi.org/10.1016/j.biortech.2019.03.061.
50. Saidulu, D.; Majumder, A.; Gupta, A. K. A Systematic Review of Moving Bed Biofilm Reactor, Membrane Bioreactor, and Moving Bed Membrane Bioreactor for Wastewater Treatment: Comparison of Research Trends, Removal Mechanisms, and Performance. *J. Environ. Chem. Eng.* **2021**, *9* (5), 106112. https://doi.org/10.1016/j.jece.2021.106112.
51. Abegglen, C.; Ospelt, M.; Siegrist, H. Biological Nutrient Removal in a Small-Scale MBR Treating Household Wastewater. *Water Res.* **2008**, *42* (1–2), 338–346.

52. Abegglen, C.; Siegrist, H. Domestic Wastewater Treatment with a Small-Scale Membrane Bioreactor. *Water Sci. Technol.* **2006**, *53* (3), 69–78.
53. Frederickson, K.; Cicek, N. Performance Comparison of a Pilot-Scale Membrane Bioreactor and a Full-Scale Sequencing Batch Reactor with Sand Filtration: Treatment of Low Strength Wastewater from a Northern Canadian Aboriginal Community. *Water Sci. Technol.* **2005**, *51* (6–7), 483–490.
54. Cote, P.; Thompson, D. Wastewater Treatment Using Membranes: The North American Experience. *Water Sci. Technol.* **2000**, *41* (10–11), 209–215.
55. Xiao, K.; Liang, S.; Wang, X.; Chen, C.; Huang, X. Current State and Challenges of Full-Scale Membrane Bioreactor Applications: A Critical Review. *Bioresour. Technol.* **2019**, *271*, 473–481. https://doi.org/10.1016/j.biortech.2018.09.061.
56. Zhang, R.; Liu, Y.; He, M.; Su, Y.; Zhao, X.; Elimelech, M.; Jiang, Z. Antifouling Membranes for Sustainable Water Purification: Strategies and Mechanisms. *Chem. Soc. Rev.* **2016**, *45* (21), 5888–5924.
57. Liu, Q.; Ren, J.; Lu, Y.; Zhang, X.; Roddick, F. A.; Fan, L.; Wang, Y.; Yu, H.; Yao, P. A Review of the Current In-Situ Fouling Control Strategies in MBR: Biological versus Physicochemical. *J. Ind. Eng. Chem.* **2021**, *98*, 42–59. https://doi.org/10.1016/j.jiec.2021.03.042.
58. Meng, F.; Chae, S.-R.; Drews, A.; Kraume, M.; Shin, H.-S.; Yang, F. Recent Advances in Membrane Bioreactors (MBRs): Membrane Fouling and Membrane Material. *Water Res.* **2009**, *43* (6), 1489–1512. https://doi.org/10.1016/j.watres.2008.12.044.
59. Rajendran, D. S.; Devi, E. G.; Subikshaa, V. S.; Sethi, P.; Patil, A.; Chakraborty, A.; Venkataraman, S.; Kumar, V. V. Recent Advances in Various Cleaning Strategies to Control Membrane Fouling: A Comprehensive Review. *Clean Technol. Environ. Policy* **2025**, *27* (2), 649–664. https://doi.org/10.1007/s10098-024-03000-z.
60. Wang, Z.; Ma, J.; Tang, C. Y.; Kimura, K.; Wang, Q.; Han, X. Membrane Cleaning in Membrane Bioreactors: A Review. *J. Membr. Sci.* **2014**, *468*, 276–307. https://doi.org/10.1016/j.memsci.2014.05.060.
61. Banerjee, S.; Manoharan, S.; Ravikumar, N.; Mahata, C. Polymeric Membrane Bioreactors in Water Treatment: A Path to Sustainability. *Iran. Polym. J.* **2025**. https://doi.org/10.1007/s13726-025-01528-5.
62. Qin, L.; Zhang, Y.; Xu, Z.; Zhang, G. Advanced Membrane Bioreactors Systems: New Materials and Hybrid Process Design. *Bioresour. Technol.* **2018**, *269*, 476–488. https://doi.org/10.1016/j.biortech.2018.08.062.
63. Mofijur, M.; Hasan, M. M.; Sultana, S.; Kabir, Z.; Djavanroodi, F.; Ahmed, S. F.; Jahirul, M. I.; Badruddin, I. A.; Khan, T. M. Y. Advancements in Algal Membrane Bioreactors: Overcoming Obstacles and Harnessing Potential for Eliminating Hazardous Pollutants from Wastewater. *Chemosphere* **2023**, *336*, 139291. https://doi.org/10.1016/j.chemosphere.2023.139291.
64. Siagian, U. W. R.; Khoiruddin, K.; Ting, Y. P.; Boopathy, R.; Wenten, I. G. Advances in Membrane Bioreactor: High Performance and Antifouling Configurations. *Curr. Pollut. Rep.* **2022**, *8* (2), 98–112.
65. Khoiruddin, K.; Boopathy, R.; Kawi, S.; Wenten, I. G. Towards Next-Generation Membrane Bioreactors: Innovations, Challenges, and Future Directions. *Curr. Pollut. Rep.* **2025**, *11* (1), 15. https://doi.org/10.1007/s40726-025-00345-x.
66. Turan, M. Membrane Fouling and Control Approaches in Membrane Bioreactor Systems: A Review. *Water. Air. Soil Pollut.* **2025**, *236* (3), 1–40.
67. Liu, J.; Liu, L.; Gao, B.; Yang, F. Integration of Bio-Electrochemical Cell in Membrane Bioreactor for Membrane Cathode Fouling Reduction through Electricity Generation. *J. Membr. Sci.* **2013**, *430*, 196–202.

68. Zhou, G.; Zhou, Y.; Zhou, G.; Lu, L.; Wan, X.; Shi, H. Assessment of a Novel Overflow-Type Electrochemical Membrane Bioreactor (EMBR) for Wastewater Treatment, Energy Recovery and Membrane Fouling Mitigation. *Bioresour. Technol.* **2015**, *196*, 648–655.
69. Li, Y.; Sun, J.; Liu, L.; Yang, F. A Composite Cathode Membrane with CoFe 2 O 4–rGO/PVDF on Carbon Fiber Cloth: Synthesis and Performance in a Photocatalysis-Assisted MFC-MBR System. *Environ. Sci. Nano* **2017**, *4* (2), 335–345.
70. Xiao, D.; Tang, C. Y.; Zhang, J.; Lay, W. C.; Wang, R.; Fane, A. G. Modeling Salt Accumulation in Osmotic Membrane Bioreactors: Implications for FO Membrane Selection and System Operation. *J. Membr. Sci.* **2011**, *366* (1–2), 314–324.
71. Luo, W.; Xie, M.; Song, X.; Guo, W.; Ngo, H. H.; Zhou, J. L.; Nghiem, L. D. Biomimetic Aquaporin Membranes for Osmotic Membrane Bioreactors: Membrane Performance and Contaminant Removal. *Bioresour. Technol.* **2018**, *249*, 62–68.
72. Wu, J.; Huang, X. Use of Ozonation to Mitigate Fouling in a Long-Term Membrane Bioreactor. *Bioresour. Technol.* **2010**, *101* (15), 6019–6027.
73. Sánchez Pérez, J. A.; Román Sánchez, I. M.; Carra, I.; Cabrera Reina, A.; Casas López, J. L.; Malato, S. Economic Evaluation of a Combined Photo-Fenton/MBR Process Using Pesticides as Model Pollutant. Factors Affecting Costs. *J. Hazard. Mater.* **2013**, *244–245*, 195–203. https://doi.org/10.1016/j.jhazmat.2012.11.015.
74. Vaccari, M.; Abbà, A.; Bertanza, G.; Collivignarelli, C. An Evidence-Based Survey on Full-Scale Membrane Biological Reactors: Main Technical Features and Operational Aspects. *Appl. Sci.* **2022**, *12* (13), 6559. https://doi.org/10.3390/app12136559.
75. Andersson, S. L.; Westling, K.; Andersson, S.; Karlsson, J.; Narongin, M.; Carranza Munoz, A.; Bornold, N.; Baresel, C. Long Term Trials with Membrane Bioreactor for Enhanced Wastewater Treatment Coupled with Compact Sludge Treatment-Pilot Henriksdal 2040, Results from 2020. **2021**.
76. Andersson, S. L.; Baresel, C.; Andersson, S.; Westling, K.; Eriksson, M.; Munoz, A. C.; Persson, G.; Narongin-Fujikawa, M.; Johansson, K.; Rydberg, T. Chemical-Saving Potential for Membrane Bioreactor (MBR) Processes Based on Long-Term Pilot Trials. *Membranes* **2024**, *14* (6), 126. https://doi.org/10.3390/membranes14060126.
77. Wen, Q.; Tutuka, C.; Keegan, A.; Jin, B. Fate of Pathogenic Microorganisms and Indicators in Secondary Activated Sludge Wastewater Treatment Plants. *J. Environ. Manage.* **2009**, *90* (3), 1442–1447.
78. Bej, S.; Swain, S.; Bishoyi, A. K.; Mandhata, C. P.; Sahoo, C. R.; Padhy, R. N. Wastewater-Associated Infections: A Public Health Concern. *Water. Air. Soil Pollut.* **2023**, *234* (7), 444.
79. Black & Veatch Corporation. *White's Handbook of Chlorination and Alternative Disinfectants*, 1st ed.; Wiley, 2009. https://doi.org/10.1002/9780470561331.
80. Mazhar, M. A.; Khan, N. A.; Ahmed, S.; Khan, A. H.; Hussain, A.; Rahisuddin; Changani, F.; Yousefi, M.; Ahmadi, S.; Vambol, V. Chlorination Disinfection By-Products in Municipal Drinking Water – A Review. *J. Clean. Prod.* **2020**, *273*, 123159. https://doi.org/10.1016/j.jclepro.2020.123159.
81. Xu, P.; Janex, M.-L.; Savoye, P.; Cockx, A.; Lazarova, V. Wastewater Disinfection by Ozone: Main Parameters for Process Design. *Water Res.* **2002**, *36* (4), 1043–1055.
82. Von Gunten, U. Ozonation of Drinking Water: Part I. Oxidation Kinetics and Product Formation. *Water Res.* **2003**, *37* (7), 1443–1467.
83. Hijnen, W.; Beerendonk, E.; Medema, G. J. Inactivation Credit of UV Radiation for Viruses, Bacteria and Protozoan (Oo) Cysts in Water: A Review. *Water Res.* **2006**, *40* (1), 3–22.
84. Sinha, R. P.; Häder, D.-P. UV-Induced DNA Damage and Repair: A Review. *Photochem. Photobiol. Sci.* **2002**, *1* (4), 225–236. https://doi.org/10.1039/b201230h.

85. Anastasi, E. M.; Wohlsen, T. D.; Stratton, H. M.; Katouli, M. Survival of Escherichia Coli in Two Sewage Treatment Plants Using UV Irradiation and Chlorination for Disinfection. *Water Res.* **2013**, *47* (17), 6670–6679. https://doi.org/10.1016/j.watres.2013.09.008.
86. Song, K.; Mohseni, M.; Taghipour, F. Application of Ultraviolet Light-Emitting Diodes (UV-LEDs) for Water Disinfection: A Review. *Water Res.* **2016**, *94*, 341–349.
87. Zhang, G.; Li, W.; Chen, S.; Zhou, W.; Chen, J. Problems of Conventional Disinfection and New Sterilization Methods for Antibiotic Resistance Control. *Chemosphere* **2020**, *254*, 126831. https://doi.org/10.1016/j.chemosphere.2020.126831.
88. Rauch, K. D.; MacIsaac, S. A.; Reid, B.; Mullin, T. J.; Atkinson, A. J.; Pimentel, A. L.; Stoddart, A. K.; Linden, K. G.; Gagnon, G. A. A Critical Review of Ultra-Violet Light Emitting Diodes as a One Water Disinfection Technology. *Water Res. X* **2024**, *25*, 100271. https://doi.org/10.1016/j.wroa.2024.100271.

Tertiary and Advanced Treatment Technologies

4

Abstract

This chapter gives an overview of tertiary treatment processes, which are advanced methods used after primary and secondary treatment to further improve wastewater quality. The objectives of tertiary treatment, including nutrient removal, pathogen reduction, and removal of residual suspended solids, trace organics, and emerging contaminants are outlined in this chapter. Key technologies discussed include filtration (sand, dual-media, membrane systems), disinfection (chlorination, ozonation, ultraviolet), nutrient removal methods (biological and chemical), and advanced oxidation processes. The chapter also addresses process selection criteria, operational considerations, and performance efficiencies. Emphasis is placed on the role of tertiary treatment in meeting stringent regulatory standards, protecting sensitive receiving waters, and enabling water reuse. Challenges such as cost, energy use, and sludge management are also highlighted, alongside trends toward integrated, sustainable treatment solutions.

Tertiary wastewater treatment, or advanced or polishing treatment, is the final and most refined stage in the treatment process. It is designed to further improve effluent quality beyond secondary treatment, particularly when stringent discharge regulations must be met or when the treated water is intended for reuse. This stage focuses on removing residual suspended solids, nutrients like nitrogen and phosphorus, pathogens, heavy metals, and emerging contaminants such as pharmaceuticals, microplastics, and antibiotic-resistant bacteria. Initial treatment may render water suitable for non-potable uses like irrigation, while complete disinfection during tertiary treatment can make it safe for potable reuse [1].

A. H. Bari et al., *Wastewater Treatment*, Synthesis Lectures on Sustainable Development,
https://doi.org/10.1007/978-3-032-07585-7_4

Various processes are employed in this phase, including filtration using sand or membrane systems, ion exchange, and disinfection methods such as ultraviolet (UV) irradiation, ozone treatment, and chlorination. Advanced oxidation processes are used for contaminants of emerging concern (CECs) and pathogens that resist conventional treatment. These processes generate highly reactive radicals (e.g., •OH, H_2O_2, O_3) capable of rapidly degrading persistent compounds and inactivating pathogens. Though tertiary treatment adds complexity and cost, it is crucial for safeguarding public health, protecting aquatic ecosystems, and enabling sustainable water reuse.

4.1 Granular Media Filtration

Granular media filtration is a widely applied physical process in water and wastewater treatment to remove suspended solids and particulate contaminants through porous filter beds. The filtration medium typically consists of granular materials, sometimes locally available, such as sand, gravel, anthracite, charcoal, pumice, zeolite, or combinations thereof [2]. The basic principle revolves around the physical straining and adsorption of particles as water flows through the granular matrix. This process plays a vital role as a tertiary treatment step to ensure the removal of residual suspended matter, thereby improving effluent clarity and quality. The effectiveness of filtration depends on factors such as media grain size, bed depth, flow rate, and the hydraulic loading conditions. There are three main configurations of granular media filters: rapid sand filters, dual media filters, and slow sand filters. The configuration employed generally depends on the nature of influent water and the required treatment standards.

4.1.1 Rapid Sand Filtration

Rapid sand filtration is one of the most commonly used water clarification methods in municipal and industrial applications. It involves using coarse granular media, typically consisting of sand supported by a gravel bed, through which water is passed at high flow rates of 4–21 $m^3/m^2/h$. The process primarily removes suspended solids by physical straining and sedimentation within the void spaces between sand particles. Unlike slow sand filters, rapid filters rely on frequent backwashing to restore performance, as the accumulation of solids can lead to clogging. This backwashing is typically performed using upward-flowing water and sometimes with air scouring to agitate and remove the trapped solids [3].

The filter bed generally has a depth of 0.6–0.75 m, and the adequate size of the sand grains ranges from 0.35 to 0.6 mm, with a uniformity coefficient of less than 1.74. As particles accumulate on the top layer, the filter exhibits increasing head loss, which triggers cleaning. Due to their high filtration rates and compact size, rapid sand filters are well-suited for large-scale installations with limited land availability. However, their performance is sensitive to influent turbidity, and they often require prior coagulation, flocculation, and sedimentation for optimal efficiency. While highly efficient, rapid sand filters may not be as effective in removing microorganisms compared to slower filtration systems.

4.1.2 Dual Media Filters

Dual media filters represent an advancement over conventional single-media rapid sand filters by employing a stratified filter bed of two different types of media, typically a coarse layer of anthracite overlying a finer sand layer. This configuration allows for deeper penetration of suspended solids into the filter bed, resulting in improved filtration efficiency and extended filter runs [3]. The anthracite media, with a lower specific gravity and larger grain size (typically 0.8–1.5 mm), rests above the denser sand layer due to fluidization dynamics during backwashing. The sand media usually has a size of 0.4–0.6 mm and a depth similar to that in rapid sand filters.

The main advantage of dual media filtration lies in maintaining low head loss while achieving higher solids loading before requiring backwashing. It is particularly effective in handling high turbidity water and extensively used in tertiary wastewater treatment and drinking water purification. Compared to single media filters, dual media configurations can double the run time between backwashing and improve particle removal without increasing energy consumption or footprint. Studies have demonstrated that dual media filters show better retention of finer particles and provide more uniform filter exhaustion profiles, translating into better operational control and lower maintenance needs.

Backwashing dual media filters is more complex due to the differing densities of the media layers. The process must be carefully designed to avoid media mixing or loss, requiring optimization of flow rates and air scour conditions. Despite these operational complexities, the enhanced performance makes dual media filters a preferred choice for applications requiring consistent high-quality effluent [4].

4.1.3 Slow Sand Filters

Slow sand filtration is among the oldest and most biologically oriented techniques suitable for decentralized or rural water treatment systems. Unlike rapid and dual media filters, slow sand filters operate at low hydraulic loading rates (0.1–0.4 $m^3/m^2/h$), allowing for the gradual percolation of water through a fine-grained sand bed. The typical sand size ranges from 0.15 to 0.35 mm, with a bed depth of approximately 0.9–1.2 m^3. A key distinguishing feature of slow sand filtration is forming a biologically active layer called the *schmutzdecke*, which develops on the sand surface and plays a vital role in removing pathogens and organic matter. This layer consists of microorganisms such as bacteria, protozoa, and algae that actively consume organic compounds and trap pathogens, enabling effective microbiological purification. The *schmutzdecke* also forms a reddish-brown sticky film composed of decomposing organics, iron, manganese, and silica, which functions as a fine filtration layer, aiding in the removal of turbidity and the reduction of taste, odour, and colour from raw water [5]. This makes slow sand filters particularly effective in pathogen removal, including Giardia and Cryptosporidium, even without chemical disinfection. The filtration mechanism in these systems is a combination of physical straining, adsorption, and biological degradation.

One of the significant advantages of slow sand filtration is its simplicity and reliability, making it ideal for communities with limited access to technical infrastructure or financial resources. Maintenance typically involves the periodic scraping and removal of the top clogged sand layer, which is then replaced with fresh or cleaned media. However, the slow filtration rate and the large land area required are significant drawbacks, limiting their applicability in high-demand urban settings [4].

Additionally, slow sand filters are sensitive to variations in influent water quality and exceptionally high turbidity, which can rapidly clog the top layer and reduce performance. Pre-treatment steps such as sedimentation or roughing filters may be necessary to ensure sustainable operation. Despite these limitations, slow sand filtration remains a sustainable and cost-effective option for small-scale water treatment with proven efficacy in improving public health outcomes.

4.2 Membrane Filtration Processes

Membrane filtration is a physical separation technique that uses semi-permeable membranes to remove suspended solids, dissolved contaminants, and pathogens from wastewater. When applied as a stand-alone process-commonly referred to as direct membrane filtration (DMF)-it operates without any integrated biological treatment, distinguishing it from MBRs, which combine biological degradation with membrane separation. This distinction makes DMF particularly suitable in scenarios where biological processes are ineffective or impractical, such as in treating low-strength wastewater or streams containing compounds that inhibit microbial activity [6].

DMF is especially advantageous in decentralized or compact treatment systems, for polishing effluent from secondary treatment, or in cases where biological treatment is either unnecessary or undesirable. It is also beneficial for treating wastewaters with variable compositions or those containing valuable components that can be recovered. Despite these advantages, membrane fouling remains a significant operational challenge, making proper pre-treatment and regular cleaning essential for maintaining system performance. Membrane filtration for wastewater treatment includes a range of processes such as pressure-driven, osmotic-driven, thermal-driven, and electrically-driven systems

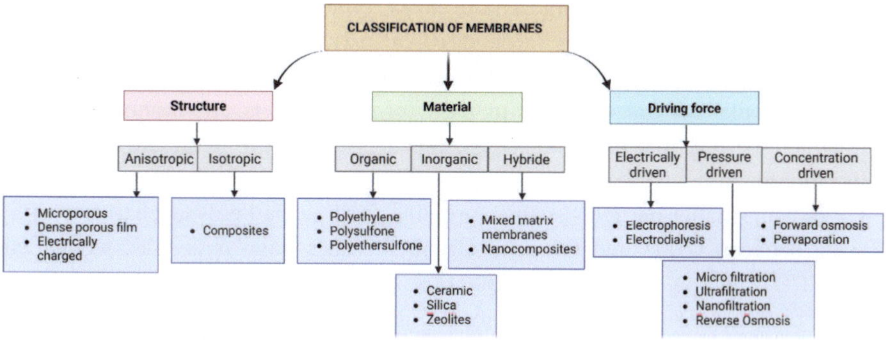

Fig. 4.1 Classification of membranes

[6]. A typical classification of membranes based on different characteristics is shown in the Fig. 4.1.

4.2.1 Pressure-Driven Membrane Technologies

Membrane filtration by pressure-driven mechanism is widely used in water and wastewater treatment, with applications ranging from solid–liquid separation to removing dissolved contaminants. The four major types-microfiltration (MF), ultrafiltration (UF), nanofiltration (NF), and reverse osmosis (RO) are distinguished primarily by pore size, separation mechanism, and the size of particles they retain, are summarized in the Fig. 4.2.

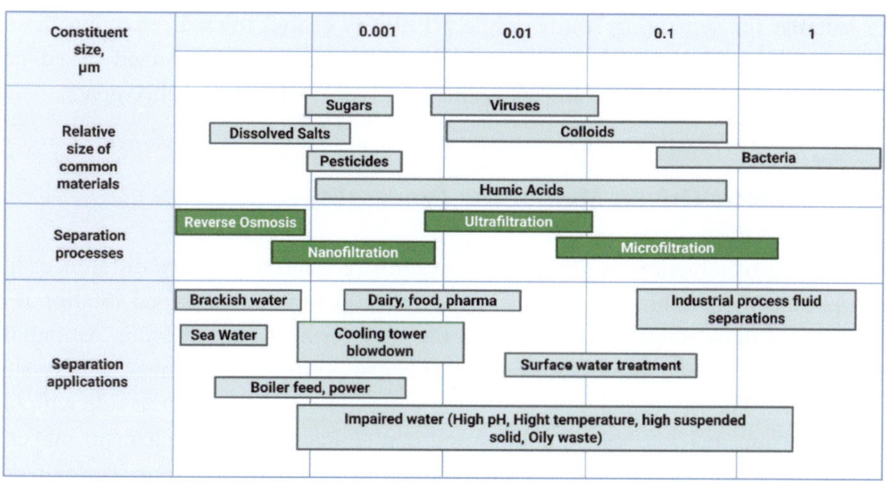

Fig. 4.2 Characteristics of different membrane processes

(i) MF membranes have the largest pore sizes, typically ranging from 0.1 to 10 μm, and operate at low pressures (around 10–50 psi) [7]. They are designed to remove suspended solids, bacteria, and large colloids, but allow most dissolved salts and low molecular weight organics to pass through. MF is commonly used for pre-treatment before finer filtration stages, clarification of beverages, and sterile filtration in bioprocessing.
(ii) UF targets smaller particles such as proteins, viruses, and polysaccharides, with pore sizes between 0.002 and 0.1 μm. UF membranes have molecular weight cutoffs (MWCO) ranging from 1,000 to 300,000 Daltons, and typically operate at 15–100 psi [7]. UF acts as an effective barrier for viruses and colloids and is used extensively in the food industry (e.g., whey protein concentration) and in tertiary wastewater treatment and feedwater polishing before RO.
(iii) NF bridges the gap between UF and RO. It uses nonporous (dense) membranes and operates on the solution-diffusion principle. NF selectively removes multivalent ions, small organics, and divalent salts, while allowing monovalent ions like sodium and chloride to pass through to some extent. NF membranes typically reject 90–99% of divalent salts, and operate at 50–225 psi [7]. They are often applied in softening hard water, separating sugars, and recovering antibiotics.
(iv) RO uses the tightest membranes, also nonporous, to remove nearly all dissolved salts, organic molecules, and even small ions. RO membranes can reject 95–99.8% of total dissolved solids, requiring pressures of 100–1,000 psi [7]. RO is widely used in seawater desalination, boiler feed water production, and industrial ultrapure water generation.

While MF and UF operate via size exclusion through microporous membranes, NF and RO rely on solution-diffusion across dense membranes. This difference makes MF/UF more suitable for separating solids, while NF/RO is geared toward removing dissolved species. Each membrane process has unique advantages and is selected based on the specific separation target, operating pressure tolerance, and water quality goals.

4.2.2 Osmotic-Driven Membrane Processes

Osmotically-driven membrane (OM) processes utilize osmotic pressure differences across a semi-permeable membrane to induce water transport from a dilute feed solution (FS) to a concentrated draw solution (DS), without the need for hydraulic pressure. Among these, forward osmosis (FO) and pressure-retarded osmosis (PRO) are the most widely studied configurations. The key principle is that water naturally moves toward the side with higher osmotic pressure, leaving behind solutes and contaminants. This mechanism makes OM processes attractive for applications such as wastewater concentration, seawater desalination, and energy recovery. The performance of OM membranes largely depends on parameters such as water permeability (A), solute permeability (B), and the structural parameter (S), which reflects the extent of internal concentration polarization (ICP). Minimizing ICP and maximizing membrane selectivity are essential for achieving high water flux and effective solute rejection. Compared to pressure-driven membranes like reverse osmosis (RO), OM membranes typically operate under milder conditions, making them suitable for energy-efficient applications.

OM technologies offer several advantages, including low energy consumption (especially in FO), reduced fouling tendencies, and the ability to treat complex wastewater streams. However, key limitations include low water flux due to ICP, membrane scaling, and the need for DS regeneration, which can complicate system operation and cost-efficiency. To address these challenges, membrane design strategies focus on optimizing the support layer's porosity and structural features and improving the active layer's hydrophilicity and selectivity. An important design aspect of FO membranes is their orientation, primarily categorized into Active Layer facing Feed Solution (AL-FS) and Active Layer facing Draw Solution (AL-DS) [8] configurations (Fig. 4.3). In AL-FS orientation, the feed solution contacts the active (dense) layer, while the draw solution interfaces with the support layer. This setup is commonly preferred in practical applications because it minimizes fouling on the membrane's support side and ensures high contaminant rejection. However, it is more vulnerable to ICP in the support layer, which can significantly reduce the effective osmotic driving force.

In contrast, the AL-DS orientation places the active layer in contact with the draw solution and the support layer in contact with the feed solution. This arrangement can lead to lower ICP and water flux under controlled conditions. However, it also exposes the porous support layer to fouling-prone feedwater, making cleaning more difficult and affecting long-term durability. Therefore, the choice between AL-FS and AL-DS must consider the specific application's balance between flux performance and fouling risk. In recent years, several promising membranes have been developed for OM processes. Hydration Technology Innovations commercialized early FO membranes as cellulose triacetate (CTA) membranes [9], which demonstrated superior FO water flux compared to commercial RO

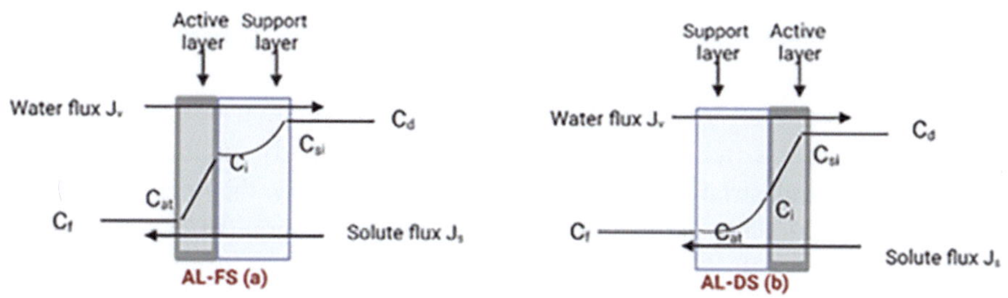

Fig. 4.3 Concentration profile in membrane cross-section. **a** AL-FS orientation; and **b** AL-DS orientation

membranes and have been widely used in OM studies. Toyobo10 later developed hollow fiber variants. Thin-film composite (TFC) FO membranes were introduced by companies like HTI, Oasys Water, and Aquaporin [10–12]. Lab-scale innovations include integral asymmetric [13], TFC [14], and dual-layer membranes [15], varying from RO-like [14], NF-like [16], to UF-like [17] structures.

Overall, OM membranes represent a promising alternative to conventional filtration processes in applications where high solute rejection, low fouling, and energy efficiency are key priorities. However, their broader adoption depends on further membrane development to overcome structural and operational limitations, especially under high-ICP and variable feed conditions.

4.2.3 Thermal-Driven Membrane Processes

Thermal-driven membrane processes, such as membrane distillation (MD) and pervaporation (PV), have emerged as promising alternatives to conventional membrane technologies due to their ability to operate at low pressures and utilize low-grade heat. These processes rely on a temperature-induced vapor pressure gradient across a membrane to drive the separation of components. In both cases, feed water is heated to induce evaporation of volatile components, passing through a membrane and condensing on the cooler permeate side, leaving behind non-volatile impurities.

MD, discovered in the 1960s [18], induces separation via evaporation, diffusion, and condensation. A hydrophobic membrane prevents liquid water from entering its pores, allowing only vaporized molecules to pass through and condense on the permeate side. MD theoretically enables near-complete rejection of non-volatile solutes like salts and large organic molecules [19]. On the other hand, in PV proposed in 1970s [20], a dense membrane with selective sorption properties is used. The permeating component is first adsorbed, then diffuses through the membrane due to a partial vapor pressure difference, and finally desorbs on the downstream side [21].

As shown in the Fig. 4.4, MD (a) involves water vapor transfer across a hydrophobic membrane from a heated feed to a colder permeate side, whereas PV (b) relies on selective sorption and diffusion through a dense membrane, suitable for separating water or volatile organics from complex mixtures.

These processes offer several advantages. Feed concentration does not significantly affect their driving force (temperature), making them particularly suitable for high-salinity brines or complex industrial wastewaters [22–24]. Additionally, they operate at relatively mild conditions (<130 °C) and can use low-grade thermal energy from waste heat, solar collectors, or geothermal sources [24]. However, their major disadvantages include lower permeate flux than RO and high energy consumption due to the heat required for vaporization [25]. PV membranes have a permeate PV flux even lower than MD, as they have a dense selective layer. Also, both PV and MD membranes' long-term stability is very poor

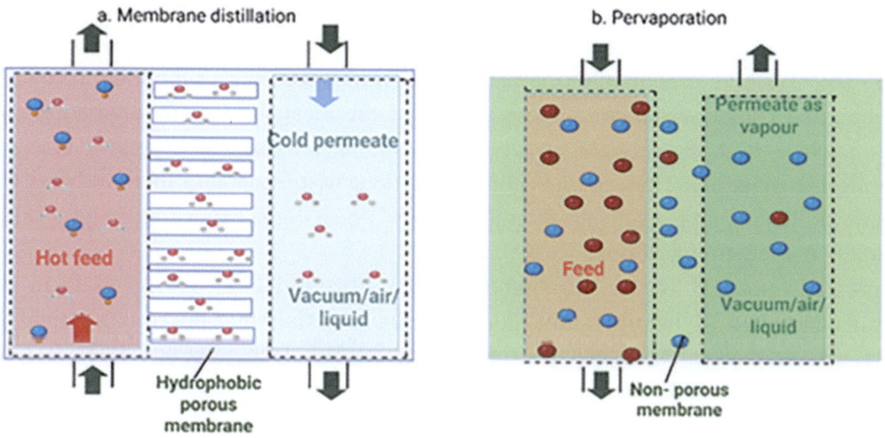

Fig. 4.4 Schematic illustration of the mechanism of **a** MD and **b** PV

[26]. Multiple design configurations of MD have been developed for various applications and have overcome these shortcomings. This includes direct contact MD (DCMD), air gap MD (AGMD), sweeping gas MD (SGMD), and vacuum MD (VMD) [27–31]. PV systems similarly use configurations resembling VMD or SGMD setups [20].

Recent studies have demonstrated the influence of operational parameters such as temperature, pH, membrane materials (PTFE, PVDF, PP), and surface properties (hydrophobicity, roughness) on MD performance [32–34]. For instance, increasing feed temperature significantly boosts permeate flux, though it may also intensify membrane fouling and wetting risks [35, 36]. Similarly, operating at higher pH levels (e.g., pH 11.5) has improved flux and reduced membrane wetting [37]. PTFE membranes showed higher permeate flux but increased wetting compared to PVDF, which offered better rejection and fouling resistance [37, 38]. Addressing fouling and wetting through proper membrane selection and cleaning strategies remains essential for long-term operation [39, 40].

4.2.4 Membrane Materials

Membrane materials are essential for the performance and efficiency of modern water purification technologies. Broadly, these materials fall into two main categories: organic (polymeric) and inorganic membranes, each offering distinct advantages based on their composition and structure [42]. Organic membranes are the most widely used and are typically made from synthetic polymers such as polyvinylidene fluoride (PVDF), polysulfone (PSU), polyethersulfone (PES), polyacrylonitrile (PAN), and cellulose acetate (CA). These polymeric materials provide flexibility in membrane design, ease of fabrication, and are compatible with a range of filtration processes such as MF, UF, NF, and

RO, making them suitable for diverse water treatment applications. In contrast, inorganic membranes-particularly ceramic-based materials like titanium dioxide (TiO_2), aluminum oxide (Al_2O_3), and zirconium oxide (ZrO_2) are known for their exceptional mechanical, thermal, and chemical stability. These properties are ideal for extreme operating conditions, such as high-temperature or highly contaminated wastewater streams. Moreover, ceramic materials like TiO_2 possess photocatalytic activity, enabling them to break down organic pollutants and inactivate microorganisms under light exposure, thereby offering an added disinfection function.

Inorganic–organic hybrid membranes are an advanced class of materials in water treatment, created by incorporating inorganic nanomaterials into polymer matrices [41]. Common additives include metal oxides (e.g., Al_2O_3, TiO_2, ZnO), metals (e.g., Cu, Ag), and carbon-based materials like graphene and carbon nanotubes. These additions improve membrane properties such as hydrophilicity, mechanical strength, permeability, solute rejection, and antifouling behavior. The performance enhancements stem from changes in membrane surface and pore structure, influenced by the thermodynamics and kinetics of membrane formation. Various fabrication methods are used to achieve this integration, including blending, phase inversion, interfacial polymerization, self-assembly, surface coating, and layer-by-layer deposition. These techniques allow precise control over material distribution and functionality, enabling membranes tailored for specific treatment needs.

4.2.5 Membrane Module

Membrane modules are engineered devices that house membranes to enable efficient separation over large surface areas. Several configurations exist to accommodate different operational needs, such as fouling control, pressure tolerance, surface area, and cost. The most common module types include plate-and-frame, tubular, spiral-wound, hollow-fiber, and capillary modules [42].

(i) The plate-and-frame module is among the earliest configurations, resembling a filter press with flat membranes sandwiched between spacers. These modules are easy to clean and allow excellent flow control, offering membrane areas of 5–20 m^2, but they are expensive and prone to leakage. With low packing density (~200–500 m^2/m^3) [43] and low pressure drop, they offer good fouling control and are used in pervaporation and electro-dialysis.

(ii) Tubular modules feature membranes cast on rigid tubes' inner or outer surface (typically 0.5–3 cm diameter), allowing high cross-flow velocities and easy cleaning. With surface areas of 1–10 m^2 per module, they are suited for ultrafiltration of high-solid or viscous feeds. They offer excellent fouling resistance, low pressure drop, and marginal suitability for high-pressure use, but are costly and bulky due to their low packing density.

(iii) The spiral-wound module consists of flat-sheet membranes rolled around a perforated central tube, forming a spiral configuration. Each spiral may have ~30 envelopes, achieving 40–60 m^2 membrane areas per module. These modules are compact, cost-effective, and suitable for RO and NF, with moderate packing density (500–1000 m^2/m^3) [43], moderate fouling control, and good high-pressure compatibility.

(iv) Hollow-fiber modules contain thousands of hair-thin fibers (ID < 0.5 mm) potted into a bundle that enables inside-out or outside-in flow. They provide incredibly high packing density (up to 5000 m^2/m^3) [43] in a compact form and are ideal for gas separation and seawater desalination. With a low cost, high-pressure resistance, and excellent space efficiency, they suffer from high pressure drop and poor fouling resistance, making cleaning difficult.

(v) Capillary modules, a variant of hollow fibers, use self-supporting fibers of 0.5–2 mm diameter, arranged in shell-tube form. They operate in inside-out mode and offer surface areas of 10–100 m^2 per module with moderate packing density (600–1200 m^2/m^3). Costing, they allow back-flushing for fouling control, have moderate pressure drop, but are restricted to low-pressure operations (4–6 bar) and are relatively fragile.

Each module type presents a trade-off among packing density, cost, fouling resistance, and pressure tolerance. Choosing the correct configuration depends on the application: liquid or gas separation, pressure limits, feed quality, and economic feasibility.

4.2.6 Applications

Membrane separation technologies have been widely applied for the removal of diverse contaminants including pharmaceuticals, dyes, heavy metals, microorganisms, pesticides, mycotoxins, and plasticizers from water. NF and RO membranes show the highest removal rates for active pharmaceutical compounds (PhACs) such as atenolol, diclofenac, and naproxen, exceeding 90% due to size exclusion and electrostatic interactions [44, 45]. Thin-film nanocomposite (TFN) membranes embedded with zeolites had rejection performance of > 90% for negatively charged PhACs [46]. Poly (vinylidene fluoride) or PVDF membranes functionalized with β-cyclodextrin achieved up to 99.9% removal of propranolol and bisphenol A [47].

In the textile and dyeing industries, NF membranes made of Polyethersulfone (PES) have demonstrated removal efficiencies ranging from 89–99.4% for dyes such as direct yellow 12, reactive green 19, and reactive blue 12 [48]. Incorporation of TiO_2 or rGO/TiO_2 nanoparticles enhances dye adsorption through increased porosity and surface reactivity [48]. Heavy metals like Cr removal using functionalized nanocomposite membranes by materials such as cellulose acetate, nanochitosan, and polyethylene glycol significantly boost removal performance beyond 90% [49]. The membrane containing 0.1 wt.% metformin-modified silica-coated Fe_3O_4 nanoparticles exhibited the copper removal efficiency of approximately 92% [50].

Membrane filtration also removes microorganisms such as bacteria and viruses from contaminated water [52]. A combination of drum filter and submerged membrane bioreactor achieved almost 5 log reduction in E. coli and Enterococci from wastewater [52] Polysulfone membrane filtration resulted in 9 log reduction of coliforms from wastewater, effectively meeting Bathing Water Directive standards [54]. Membrane filtration is an effective method for removing pesticides from contaminated water sources Nanofiltration and low desalting reverse osmosis membranes were tested for the removal of 11 non-phenylic pesticides, and the membrane with the highest desalting capacity achieved over 96.7% rejection for all pesticides except dichlorvos. The study found that pesticide rejection and adsorption on membranes were primarily influenced by hydrophobicity (log P), suggesting it as a useful predictor for membrane performance in pesticide removal [54].

4.3 Advanced Oxidation Processes

Advanced Oxidation Processes (AOPs) are a class of chemical treatment technologies that rely on the generation of highly reactive species, primarily hydroxyl radicals (•OH), to achieve the degradation of persistent organic and certain inorganic pollutants in water and wastewater systems. Initially introduced in the 1980s for potable water treatment [55, 56], AOPs have since evolved into one of the most versatile and effective methods for addressing contaminants resistant to conventional treatment. These include pharmaceuticals, pesticides, industrial chemicals, and other trace-level emerging contaminants that pose ecological and human health risks. The core principle of AOPs lies in the in situ generation of $^{\bullet}OH$ radicals in sufficient concentrations to initiate rapid, non-selective oxidative reactions, leading to the mineralization or transformation of complex pollutants into simpler, less harmful molecules.

Hydroxyl radicals are among the most potent oxidizing agents in environmental remediation. Their oxidation potential ranges from 2.8 V at acidic conditions (pH 0) to 1.95 V under fundamental conditions (pH 14), referenced against the saturated calomel electrode (SCE) [4]. This high redox potential enables $^{\bullet}OH$ radicals to react rapidly and indiscriminately with various organic substances. The reaction mechanisms include radical addition to double bonds, hydrogen abstraction from aliphatic chains, electron transfer from electron-rich species, and radical combination reactions, with rate constants typically falling within the range of 10^8 to 101^0 M^-1 s^-1 [57]. Unlike traditional oxidants such as chlorine or ozone, whih are commonly employed for contaminant removal and microbial disinfection, AOPs are designed primarily for chemical degradation. AOPs include a variety of techniques such as ozonation, Fenton and photo-Fenton reactions, UV/H_2O_2, photocatalysis, electrochemical oxidation, and combinations thereof, with newer methods also involving sulfate radical-based AOPs for enhanced pollutant degradation [58–60] (Table 4.1).

4.3.1 Ozonation

Ozonation is an oxidation process increasingly utilized in water and wastewater treatment due to its ability to degrade many refractory and toxic organic pollutants. Ozone (O_3), a triatomic form of oxygen, acts as a potent oxidant with an oxidation potential of 2.07 V, which on decomposition yields •OH radical, one of the most reactive oxidizing agents known. These characteristics enable ozonation to effectively break down persistent contaminants such as pharmaceuticals, pesticides, dyes, and industrial chemicals [69]. Moreover, ozone has excellent disinfection properties, making it suitable for microbial inactivation in municipal and industrial wastewater systems.

The mechanisms of ozonation can be broadly categorized into direct and indirect pathways. In direct ozonation, molecular ozone selectively reacts with electron-rich moieties such as double bonds, aromatic rings, and functional groups containing nitrogen, sulfur, or phosphorus. These reactions typically occur through redox transformations, cycloaddition, electrophilic substitution, and nucleophilic interactions. For instance, ozone reacts with hydroperoxide anions (HO_2^-) or superoxide (O_2^{-}•) in redox processes:

$$O_3 + HO_2^- \rightarrow O_2^{\bullet} + HO_2\bullet$$

$$O_3 + O_2^{-\bullet} \rightarrow O_3^{-\bullet} + O_2$$

Cycloaddition reactions, such as those in the Criegee mechanism, involve the attack of ozone on carbon–carbon double bonds (e.g., in olefins), forming a five-membered primary ozonide. This unstable intermediate decomposes to generate a carbonyl oxide zwitterion, which can follow different reaction pathways to yield final products such as aldehydes, ketones, or carboxylic acids [70]. A simplified version of this reaction is:

$$C = C + O_3 \rightarrow [\text{primary ozonide}] \rightarrow \text{aldehydes} + \text{ketones}$$

Indirect ozonation involves the formation of hydroxyl radicals through ozone decomposition, especially under alkaline conditions (pH > 8). These highly reactive radicals can oxidize a broad range of organic and inorganic compounds via hydrogen abstraction and radical–radical reactions [71].

The radical-based mechanism progresses through initiation, propagation, and termination stages. During initiation, ozone reacts with hydroxide ions to form a superoxide radical and hydroperoxyl radical:

Table 4.1 Comparison of different AOPs

Technique	Mechanism	Advantages	Disadvantages	Applications	References
Hydroxyl Radical-Based AOPs (•OH)	Non-selective oxidation via in situ-generated hydroxyl radicals from chemical, photolytic, or electrochemical methods	Highly reactive, decomposes broad pollutant types	High energy demand, risk of toxic intermediate formation	Industrial, hospital wastewater, textile industry	[62]
Ozone (O_3)	Oxidation through direct ozone attack and generating secondary radicals like •OH	Strong oxidant, effective for organic and inorganic contaminants, with no sludge formation	High operational cost, short ozone half-life, and potential bromate formation	Drinking water, industrial wastewater, and pharmaceuticals	[63]
UV/H_2O_2	UV photolysis of H_2O_2 forms •OH radicals	Effective for organics, no sludge, can be integrated with other methods	High UV energy cost, requires H_2O_2 dosage, and potential pre-treatment	Drinking water, groundwater remediation	[64]
Fenton (Fe^{2+}/ H_2O_2)	Fe^{2+} catalyzes H_2O_2 decomposition to form •OH radicals under acidic conditions	Cost-effective, effective for organic pollutants, and simple equipment	Sludge formation requires pH adjustment, not suitable for high-COD wastewater	Industrial wastewater, dye removal, refinery effluents	[65]
Photo-Fenton (Fe^{2+}/H_2O_2/ UV)	UV light enhances Fe^{3+} reduction and H_2O_2 breakdown, increasing •OH production	More efficient than Fenton alone, effective at lower doses	Requires a UV light source, high operational cost, and sludge generation	Recalcitrant organics, pharmaceuticals	[66]

(continued)

Table 4.1 (continued)

Technique	Mechanism	Advantages	Disadvantages	Applications	References
Ultrasound (US)	Acoustic cavitation via ultrasonic waves causes bubble collapse, generating extreme T/P and reactive species ($^{\bullet}OH$, $H^{\bullet}$, H_2O_2)	No secondary pollution, fast reaction rate, reduced dose of chemicals	High energy demand needs precise frequency, power. Excessive noise. High maintenance cost	Industrial and agricultural effluents, especially with refractory organics	[67, 68]
Hydrodynamic Cavitation (HC)	Cavitation via flow constriction (venturi/orifice), leading to bubble collapse and ROS formation	Energy-efficient, scalable, compatible with continuous systems, enhances biodegradability, and can operate without external oxidants	Sensitive to design and operating parameters, material erosion, lower radical yield than chemical AOPs; often needs hybridization (O_3, H_2O_2, UV)	Dye, pesticide, and distillery wastewater; effective as pre-treatment for biological processes	[67]
Electrochemical AOPs (EAOPs)	Electric current generates oxidants ($\bullet OH$, O_3, $Cl\bullet$) at electrodes; it avoids chemical use	Precise control, no chemical input, effective for recalcitrant pollutants	High energy consumption, electrode fouling, and costly equipment	Petrochemical, pharmaceutical, landfill leachate	[69]

$$O_3 + OH^- \rightarrow HO_4^-$$

$$HO_4^- \leftrightarrow HO_2\bullet + O_2^{-\bullet}$$

In the propagation phase, $O_2^{-}\bullet$ further reacts with ozone to produce ozonide:

$$O_3 + O_2^-\bullet \rightarrow O_3^-\bullet + O_2$$

Subsequent decomposition of ozonide anion leads to hydroxyl radical formation via intermediate hydrogen trioxide ($HO_3\bullet$):

$$O_3^{-\bullet} + H^+ \rightarrow HO_{3\bullet} \rightarrow OH\bullet + O_2$$

The termination phase involves scavenging of hydroxyl radicals by species like bicarbonate and carbonate ions, reducing oxidative potential. For example:

$$OH\bullet + HCO_3^- \rightarrow H_2O + CO_3^{-\bullet}$$

$$OH\bullet + CO_3^{2-} \rightarrow OH^- + CO_3^{-\bullet}$$

Ozonation offers several advantages over conventional treatment methods. It enables the complete mineralization of many complex and toxic organics and significantly improves the biodegradability of otherwise recalcitrant substances. Moreover, it produces fewer harmful byproducts than chlorination and is effective across a broad range of flow rates and contaminant concentrations. Its disinfection capacity also makes it a favorable alternative for pathogen control. However, several limitations restrict its stand-alone use. These include high operational costs associated with ozone generation and reactor systems, short half-life of ozone necessitating on-site production, and potential formation of toxic transformation products during partial oxidation [73]. Additionally, natural organic matter, suspended solids, and pH variations can adversely affect ozonation efficiency, which must be carefully controlled for optimal performance.

In conclusion, ozonation and its advanced variants represent a significant leap forward in treating contaminated water and wastewater. With the ability to degrade a diverse range of pollutants and improve disinfection, ozone-based technologies align well with current goals for environmental protection and public health. Nonetheless, further research and development in catalyst design, reactor optimization, and cost reduction are essential for the broader implementation of these technologies in real-world settings. Ozonation, particularly when employed in hybrid configurations, holds great promise as a cornerstone of sustainable wastewater management.

To overcome these challenges and enhance treatment efficacy, ozonation is often coupled with other physical, chemical, or biological processes, resulting in hybrid AOPs. Among these, the O_3/H_2O_2 system, known as the peroxone process, is widely used. This combination leads to enhanced generation of hydroxyl radicals due to the reaction between ozone and the hydroperoxide anion, thereby improving the breakdown of persistent pollutants:

$$HO_2^- + O_3 \rightarrow HO_2\bullet + O_2^{-\bullet} \rightarrow OH\bullet + O_2 + \text{other radicals}$$

However, the effectiveness of this system depends on the precise dosing of hydrogen peroxide, as excess H_2O_2 can act as a radical scavenger:

$$OH\bullet + H_2O_2 \rightarrow HO_2\bullet + H_2O$$

Another necessary hybrid process is the combination of ozonation with ultraviolet (UV) irradiation, known as the O_3/UV system. UV light accelerates ozone's photolytic decomposition, producing more efficient hydroxyl radicals. This method is particularly effective in degrading dye molecules and other aromatic compounds in wastewater, resulting in smaller, more biodegradable aliphatic intermediates. Reactions such as:

$$O_3 + h\nu \rightarrow O_2 + O(^1D)$$

$$O(^1D) + H_2O \rightarrow 2\,OH\bullet$$

This is fundamental to the radical generation process. Studies have shown that the O_3/UV system has a higher decolorization rate and contaminant breakdown than ozonation alone [73].

The integration of ozonation with biological treatment is also gaining traction. Pre-ozonation can break down complex and toxic compounds into simpler intermediates, thereby enhancing the effectiveness of subsequent biological treatment. Ozone pre-treatment significantly enhanced aerobic COD reduction, achieving 25 times higher biodegradation than untreated distillery wastewater [74]. This integration approach improves mineralization rates and reduces the formation of hazardous byproducts, offering a more sustainable and cost-effective solution. Similarly, catalytic ozonation, which employs transition metal oxides to accelerate ozone decomposition, is emerging as a promising strategy to boost process efficiency under milder conditions [75].

4.3.2 UV-Based Oxidation

UV-based AOPs have emerged as a powerful set of technologies for the treatment of water and wastewater contaminated with trace organic compounds. UV-based oxidation processes have been successfully applied in both pilot and full-scale systems for potable water reuse, surface water treatment, and contaminant removal in various water matrices [76]. These methods primarily rely on high-energy UV light, particularly in the UV-C range, to initiate photolytic and radical-generating reactions capable of mineralizing persistent pollutants. Given their potential to degrade a broad spectrum of pollutants and their compatibility with membrane systems and disinfection, UV-based AOPs are gaining increasing attention in potable water reuse and surface water treatment contexts [77].

The core principle behind UV-based oxidation is the photochemical generation of highly reactive radical species that oxidize organic pollutants. UV light, especially at wavelengths below 300 nm, cleaves chemical bonds either directly in the pollutants or in added oxidants, such as hydrogen peroxide (H_2O_2), ozone (O_3), peroxydisulfate (PDS), or chlorine (Cl_2), thereby initiating chain oxidation reactions. Hydroxyl radicals (•OH) are the most commonly targeted oxidants, though sulfate ($SO_4^{\bullet-}$) and chlorine radicals (Cl•, $Cl_2^{\bullet-}$) also play important roles depending on the chosen oxidant system [79].

Combinations of UV with Radical Promoters.

4.3.2.1 UV/H_2O_2

The most widely studied UV-based AOP process is the UV/H_2O_2, where UV irradiation cleaves H_2O_2 into two hydroxyl radicals. H_2O_2 photolysis rates are pH-dependent and increase under alkaline conditions. A key limitation of the process is the low molar extinction coefficient of H_2O_2 (18.6 M^{-1} cm^{-1} at 254 nm), which restricts light utilization, primarily when organic compounds act as inner filters [79]. Also, the UV/H_2O_2 process can face efficiency losses due to fouling or scaling of lamps and reactor surfaces. It may be less effective for wastewaters with recalcitrant organics, often requiring integration with other AOPs. Although it limits persistent toxic byproducts, transient or minor byproducts may still form, necessitating further research into their control. Despite these challenges, UV/ H_2O_2 has been effectively deployed at full scale for applications such as potable water reuse and surface water treatment [80–82].

4.3.2.2 UV/O_3

In the UV/O_3 system, UV light cleaves dissolved ozone into atomic oxygen, which reacts with water to form excited H_2O_2, subsequently decomposing into •OH [83]. Ozone's absorption at 254 nm is significantly higher than H_2O_2 ($\varepsilon = 3300$ M^{-1} cm^{-1}), but recombination effects and low quantum yield (around 0.1) limit the radical production [84]. From the photochemical point of view, as compared to H_2O_2, the absorption spectrum of O_3 offers a significantly higher absorption cross section. Although the combination of ozone and UV allows for broad reactivity and color removal, the process suffers from high energy demands due to the need for both ozone generation and UV lamps.

4.3.2.3 UV/$SO_4^{\bullet -}$

UV/$SO_4^{\bullet -}$ is a promising alternative to hydroxyl radical-based AOPs for the oxidation of organic contaminants in water, as it primarily generates sulfate radicals ($SO_4^{\bullet -}$) having strong oxidizing power and are more selective oxidants than •OH [85]. UV-activated sulfate radical generation involves the photolysis of peroxydisulfate (PDS) or peroxymonosulfate (PMS). PDS has a higher quantum yield (1.4) and molar absorption than H_2O_2, producing primarily $SO_4^{\bullet -}$ radicals, which are more selective oxidants compared to •OH. UV/$SO_4^{\bullet -}$systems have shown improved radical yields and degradation performance in lab-scale studies, although their sensitivity to water matrix constituents and higher oxidation byproducts formation potential pose constraints for widespread application.

4.3.2.4 UV/Cl_2

The UV/Cl_2 process activates chlorine species (mainly hypochlorite or chlorine dioxide) under UV light to form Cl•, $Cl_2^{\bullet +}$, and •OH radicals, all capable of oxidizing target pollutants [86]. These radicals are more selective than $^{\bullet}$OH and especially effective against electron-rich contaminants. The process performance of UV/Cl_2 is sensitive to water quality parameters like pH, alkalinity, and chlorine dose [87]. UV/chlorine processes offer higher energy efficiency than UV/ H_2O_2 or UV/persulfate, especially under acidic conditions. Due to widespread use of UV and chlorination units, they can be easily integrated into existing systems and show strong potential for effective contaminant degradation[89]. Potential formation of disinfection byproducts is a concern, particularly in the presence of halide ions and natural organic matter. Full-scale deployment of UV/Cl_2 has begun, notably at the Los Angeles Terminal Island Water Reclamation Plant [85].

4.3.2.5 Overall Advantages and Limitations UV UV-Based Methods

Advantages of UV-based AOPs include their strong disinfection capability, effective degradation of a wide range of micropollutants, and compatibility with membrane treatment systems. Their design flexibility is further enhanced by the emergence of UV-LEDs, which offer mercury-free operation, targeted wavelength emission, compact size, and rapid start-up [89]. However, the energy efficiency of current UV-LEDs remains low (<10%), making them less competitive than mercury-based lamps [89]. Additionally, low UV transmittance in secondary and tertiary wastewater effluents often limits UV-based processes due to high dissolved organic matter, which absorbs and scatters light, reducing radical generation efficiency. Excess dosages of oxidants like H_2O_2 or PMS may also lead to radical scavenging and unwanted byproduct formation.

4.3.3 Fenton and Photo-Fenton Processes

Fenton oxidation, involving reactions between H_2O_2 and iron ions (Fe^{2+} or Fe^{3+}), is a well-established AOP that has gained considerable attention for its powerful ability to degrade complex and recalcitrant organic pollutants. The process is highly effective in treating a wide range of industrial and agricultural wastewater types due to the generation of hydroxyl radicals (•OH).

The classic Fenton reaction proceeds via the catalytic decomposition of hydrogen peroxide by ferrous iron (Fe^{2+}) under acidic conditions (typically pH 2–4), leading to the generation of hydroxyl radicals:

$$Fe^{2+} + H_2O_2 + H^+ \rightarrow Fe^{3+} + H_2O + \bullet OH$$

These hydroxyl radicals attack and degrade a wide range of organic contaminants into simpler and often non-toxic products like carbon dioxide and water. To regenerate Fe^{2+} and continue the catalytic cycle, ferric iron (Fe^{3+}) must be reduced:

$$Fe^{3+} + H_2O_2 \rightarrow Fe^{2+} + CO^2 + H^+$$

However, formation of Fe^{3+} is faster than that of its consumption by a factor of about 6000, limiting efficient Fe^{2+} recycling and leading to the accumulation and eventual precipitation of Fe^{3+} as iron sludge above pH 3 [90]. To enhance regeneration and efficiency, the Photo-Fenton process utilizes UV or visible light to photo-reduce Fe^{3+} back to Fe^{2+} while simultaneously producing more $^{\bullet}OH$ radicals:

$$Fe^{3+} + H_2O_2 + h\nu \rightarrow Fe^{2+} + H^+ + \bullet OH$$

Alternative oxidants like ozone (O_3) and even oxygen (O_2) have been evaluated to replace or supplement H_2O_2, with Fe^{2+}/O_3/UV systems showing promise for mineralizing highly colored wastewater, though they involve slower kinetics and more complex, costly setups.

The core strength of the Fenton process lies in its ability to generate highly reactive hydroxyl radicals capable of degrading both biodegradable and refractory organics. Operating at ambient conditions and with relatively simple infrastructure, Fenton systems are cost-effective and do not require complex equipment. However, limitations include a narrow optimal pH (2–4), requiring acid addition, and the generation of problematic iron sludge. Advanced Fenton systems, such as Electro-Fenton 83, Ultrasound-assisted Fenton84, Microbial electrolysis–Fenton 8,5, etc., have been developed to overcome these issues. These advancements have broadened the scope and sustainability of Fenton-based AOPs across industrial, pharmaceutical, and agricultural wastewater treatment.

4.3.4 Cavitation-Based AOPs

Among various AOPs, cavitation-based processes (acoustic and hydrodynamic cavitation) are gaining significant attention for their effectiveness, energy efficiency, and potential scalability. Cavitation involves the formation, growth, and implosive collapse of microbubbles in a liquid medium, resulting in localized extreme conditions of temperature and pressure that trigger chemical reactions. Cavitation has been effectively used in particulate processes like crystallization [91], particle breakage [92], emulsification, etc. [93]. Cavitation can also be effective for industrial and agricultural wastewater treatment, where traditional methods fail to fully degrade persistent organic compounds.

4.3.4.1 Ultrasonic-Based AOPs

In cavitation-based systems, radical generation is driven by the violent collapse of vapor cavities. In acoustic or ultrasonic (US) cavitation, ultrasonic waves induce microbubble collapse, leading to temperatures up to 5000 K and pressures around 1000 atm. This extreme environment enables homolytic cleavage of water molecules, •OH, H•, H_2O_2, and other reactive oxygen species. The efficiency of cavitation can be significantly improved when combined with chemical or photochemical processes. An advanced application is sonophotocatalysis, which integrates ultrasound and UV irradiation with catalysts like TiO_2 or ZnO. This method benefits from both acoustic cavitation and photon-induced charge separation, enhancing reactive oxygen species (ROS) generation and degradation kinetics. Wei et al. [94] demonstrated its efficacy in degrading acid orange 7 dye, especially with added H_2O_2. Similarly, Razavi [95] reported 96.7% removal of linear alkylbenzene sulfonate within 24 min using a combined UV/US system. Process parameters such as pH, contact time, and pollutant concentration were shown to be critical.

Alfonso-Muniozguren et al. (2021) further highlighted the scalability and sustainability of sonochemical processes for treating industrial and agricultural wastewaters, especially those containing refractory pollutants. The US-based wastewater treatment process's performance depends on frequency, power input, treatment time, temperature, pH, pollutant concentration, and reactor geometry. Integrating UV and ultrasound allows more profound pollutant disruption and broader ROS generation, while catalysts like TiO_2 or ZnO amplify treatment efficiency. Despite higher energy demands and the need for optimized catalyst recovery, sonophotocatalytic systems reduce chemical usage and adapt well to diverse wastewater types.

4.3.4.2 Hydrodynamic Cavitation-Based AOPs

Hydrodynamic cavitation (HC) induces cavitation through flow constriction via orifices, venturis, or nozzles, providing a more scalable and energy-efficient approach than acoustic methods. The collapse of cavitation bubbles generates reactive radical species that initiate the oxidative breakdown of complex organic pollutants into smaller, more biodegradable intermediates. When combined with oxidants such as hydrogen peroxide or ozone, HC shows significantly enhanced performance. Biodegradability in pesticide contaminated wastewater using HC combined with H_2O_2 or ozone improved degradation with higher rates [96]. HC is especially valued for its low energy requirements, compatibility with continuous-flow systems, and potential to operate without external oxidants, thus minimizing secondary pollution and operational costs. Its ability to improve the biodegradability of effluents makes it a strong pre-treatment option for subsequent biological processes. Optimizing reactor geometry and operating parameters such as pressure drop, flow rate, and temperature has significantly influenced HC efficiency [97, 98]. Successful degradation of reactive dyes like acid red 88 and reactive red 120 is reported using optimized HC reactors. In addition to dyes, HC systems have proven effective for treating pesticide-laden effluents, achieving substantial reductions in COD and toxicity [96].

However, HC also presents certain challenges. Performance is susceptible to system design; improper configuration can result in energy inefficiency or inadequate cavitation. Long-term use may lead to material erosion due to intense bubble collapse. Furthermore, the radical yield in HC alone may be lower than that of chemical AOPs, often requiring hybridization with agents like ozone, UV, or H_2O_2 for improved treatment outcomes. Overall, HC-based systems serve as stand-alone treatment options and valuable components in multi-stage treatment frameworks, offering efficient pre-treatment or polishing capabilities across diverse industrial wastewater streams.

Cavitation-based AOPs offer up-and-coming solutions for the degradation of complex organic pollutants in industrial and agricultural wastewater. When properly optimized and integrated with other AOPs such as UV irradiation, hydrogen peroxide, and catalysts, cavitation systems can achieve superior degradation efficiencies, enhance biodegradability, and support sustainable, chemical-minimized water treatment practices. As energy-efficient and environmentally friendly alternatives, cavitation-based AOPs are well-positioned for expanded application in future wastewater treatment infrastructure.

4.3.5 Sulfate Radical-Based AOPs

Sulfate radical-based AOPs (SR-AOPs) have emerged as a powerful class of water and wastewater treatment technologies, particularly effective in the degradation of persistent organic pollutants. Traditionally, AOPs focused on the generation of hydroxyl radicals (•OH) due to their high oxidation potential and non-selective reactivity. However, since the mid-2000s, sulfate radicals ($SO_4^{\bullet -}$) have gained considerable attention as a viable and often complementary alternative to hydroxyl radical-based systems. Sulfate radicals offer comparable oxidation potential and often demonstrate greater stability and selectivity under various environmental conditions, particularly at neutral and alkaline pH levels. The generation of sulfate radicals typically involves the activation of strong oxidants such as peroxymonosulfate (PMS) and peroxydisulfate (PDS). These oxidants can be activated using various strategies, including transition metal catalysts (e.g., Fe^{2+}, Co^{2+}, Mn^{2+}), heat, UV or visible light, ultrasound, electrochemical methods, and photocatalytic excitation (h^+/e^-) [99]. The activation leads to the homolytic cleavage of the O–O bonds in the oxidants, forming sulfate radicals $SO_4^{\bullet -}$.

SR-AOPs offer several advantages over conventional oxidation processes [100]. SR-AOPs are versatile and can be implemented individually or in combination with other treatment technologies to enhance degradation efficiency. Sulfate radicals have a high oxidation potential (2.5–3.1 V) and longer half-life (30–40 μs) than hydroxyl radicals, allowing better reactivity and diffusion. They are effective across a wider pH range (2–8) and show selectivity toward specific organic compounds, such as aromatic or unsaturated structures. SR-AOPs are also suitable for removing emerging micro-pollutants and integrating with membrane systems.

Despite their promise, the application of SR-AOPs is still largely limited to laboratory studies, with field-scale implementation facing several technical and environmental challenges. One major concern is the generation of toxic byproducts, particularly transformation products from target pollutants and reactions involving effluent organic matter (EfOM). In some instances, the overall toxicity of treated water has been found to increase due to the formation of more harmful intermediate compounds than the original contaminants [100]. Therefore, SR-AOPs are often proposed in combination with post-treatment methods such as activated carbon adsorption or membrane[102] filtration to reduce residual toxicity and byproduct accumulation. Additionally, high energy input for activation (e.g., UV, ultrasound) and limited full-scale application due to cost and complexity pose challenges.

4.4 Electrochemical Treatment Technologies

Electrochemical technologies have emerged as promising methods for the treatment and remediation of wastewater due to their environmental compatibility, versatility, high energy efficiency, and operational simplicity. These processes utilize electric current to drive redox reactions that transform pollutants into less harmful substances or facilitate removal. The primary advantage is using electrons as clean reagents, eliminating the need for added chemicals. Electrochemical techniques can act as stand-alone treatment methods or be integrated with other technologies for enhanced performance and sustainability [102].

4.4.1 Electrochemical Oxidation

Electrochemical oxidation (EO) involves the degradation of contaminants at the anode surface. It can occur either via direct oxidation, where pollutants are oxidized at the electrode surface, or indirect oxidation, where reactive species generated at the electrode (such as hydroxyl radicals or active chlorine) interact with pollutants in solution. The fundamental oxidation mechanism involves physisorbed hydroxyl radicals (•OH) formed at the anode surface, which possess a high oxidation potential ($E^\circ = 2.8$ V vs SHE) [102] and can oxidize organic pollutants to carbon dioxide and water.

The electrode material significantly influences the oxidation efficiency. "Non-active" anodes like boron-doped diamond (BDD), PbO_2, or SnO_2 promote complete oxidation, while "active" anodes like IrO_2 or RuO_2 favor partial and selective oxidation. Advanced EO processes like electro-Fenton [103], photoelectro-Fenton [104], and sonoelectro-Fenton [105] involve synergistic enhancement by combining EO with other AOPs to increase radical generation and pollutant removal. The electro-Fenton method avoids Fe^{3+} accumulation and thus reduces iron sludge production. Advantages include high efficiency for non-biodegradable pollutants and low chemical usage. However, there are certain limitations like high energy demand and potential electrode fouling, especially during long-term operation. Applications span the treatment of dyes, pharmaceuticals, endocrine disruptors, and industrial effluents [102].

4.4.2 Electrochemical Reduction

Electrochemical reduction processes are effective for the removal of inorganic pollutants such as nitrates and heavy metals, as well as halogenated organic compounds [102]. In cathodic electrochemical dechlorination, pollutants like chlorinated organic compounds (COCs) are reductively dehalogenated using cathodes made from materials like palladium or platinum. Palladium, in particular, exhibits high catalytic activity due to its hydrogen adsorption properties, facilitating hydrodehalogenation reactions [106]. In electrochemical denitrification, nitrate is reduced to nitrogen gas or ammonium ions, depending on the cathode material. Transition metal electrodes such as Ruthenium, Rhodium, Iridium, and Palladium demonstrate varying selectivity and efficiency for nitrate-to-nitrite or nitrate-to-ammonia conversion. Studies show that nitrate reduction follows Langmuir–Hinshelwood kinetics and that the selectivity toward nitrogen gas increases with nitrate concentration [107].

Electrodeposition is another reductive technique used for heavy metal removal. Metals like copper, silver, and nickel can be deposited on the cathode surface and recovered. However, hydrogen evolution at the cathode may interfere with deposition efficiency [102]. The advantages of electrochemical reduction include selective pollutant removal, potential metal recovery, and minimal sludge production. However, challenges include energy costs, the need for selective electrode materials, and the formation of toxic intermediates.

4.4.3 Electrocoagulation

Electrocoagulation (EC) is a hybrid electrochemical-physical process wherein electrical current is applied to sacrificial metal electrodes (commonly iron or aluminum), causing the in situ generation of coagulants that destabilize suspended particles and facilitate pollutant removal[103]. The anodic dissolution produces metal hydroxide flocs that adsorb and sweep contaminants from water. Electrocoagulation (EC) offers key advantages such as in-situ coagulant generation without added chemicals, reduced sludge production, and improved pollutant removal through simultaneous coagulation and flotation. It produces non-toxic, stable flocs and is easy to automate, making it suitable for diverse wastewater types. However, EC has limitations, including electrode consumption, particularly anode wear and cathode passivation, which can reduce efficiency. Its performance is sensitive to factors like pH, current density, and electrode material, and operational costs may rise with electricity demand or frequent maintenance needs [108].

Overall, electrochemical technologies offer significant potential for sustainable and effective wastewater treatment. Electrochemical oxidation is especially effective for persistent organics, while reduction technologies are valuable for detoxifying halogenated compounds and recovering metals. Electrocoagulation provides a versatile and low-chemical alternative to conventional coagulation. Although challenges such as energy consumption, material cost, and scaling up remain, hybrid processes and advancements in electrode materials are steadily improving the feasibility of electrochemical methods in practical applications.

4.5 Adsorption and Ion Exchange

Adsorption and ion exchange processes have gained renewed attention, particularly within the circular economy framework, for their potential in resource recovery and environmental protection. These techniques offer cost-effective, eco-friendly, and versatile approaches to remove and recover valuable resources from wastewater streams. Their mechanisms, practical applications, and associated advantages and limitations make them vital components in modern wastewater management. Adsorption and ion exchange are considered together because they share many common features, including design concepts and, in some cases, operating cycles, although ion exchange operating cycles are somewhat more complex.

Adsorption is a surface phenomenon in which solutes present in wastewater adhere to the surface of a solid material called an adsorbent. This interaction can be driven by various forces, including physical interactions (such as van der Waals forces) and chemical bonding (such as ion exchange, hydrogen bonding, or covalent interaction). Adsorption can be classified as either physisorption, which is reversible and occurs due to weak interactions, or chemisorption, which involves stronger, often irreversible chemical bonds. The efficiency of adsorption depends on factors such as surface area, porosity, functional groups of the adsorbent, pH, temperature, and the presence of competing species in the solution. Most adsorption processes follow isotherm models like Langmuir or Freundlich that describe how solutes interact with the surface, and kinetic models that describe the rate of adsorption.

Conversely, ion exchange is a stoichiometric exchange of ions between a solid (ion-exchange resin or material) and a liquid phase. Ion exchangers are usually insoluble polymers or inorganic matrices containing mobile counter-ions. When the target ions in wastewater come in contact with the exchanger, a selective exchange occurs, and target ions are captured while the bound ions on the resin are released into the liquid. Electrostatic interactions, selectivity coefficients, and affinity sequences govern this process. Ion exchange is highly efficient for the removal of charged species, such as heavy metals, ammonium, nitrates, and certain organic acids. It is also suitable for both removal and recovery of valuable resources, due to the reversible nature of the process, which allows regeneration and reuse of resins.

Numerous studies have been conducted on wastewater treatment using adsorption and ion exchange, highlighting their effectiveness in removing a wide range of contaminants. Adsorption and ion exchange have found wide-ranging applications in removing diverse pollutants from wastewater, including dyes [109], heavy metals [110], nutrients [111], organic acids, and biosurfactants [112]. Adsorption and ion exchange are widely recognized for their robustness, adaptability, and sustainability in wastewater treatment. They offer several advantages, including operational simplicity, low energy requirements, and selectively removing a broad spectrum of contaminants such as heavy metals, nutrients, and organics. These processes often utilize renewable or reusable materials like activated carbon, natural clays, biosorbents, and synthetic resins, making them cost-effective and well-aligned with circular economy principles. Additionally, their modular design allows easy integration into existing treatment systems and flexibility across varying wastewater compositions.

Ion exchange specifically enables precise removal of target ions by exchanging them with desirable ions like H^+ or OH^- from the resin. Regeneration of saturated resins is typically done using strong acids or bases, allowing for repeated use. Although continuous regeneration systems exist, most are operated intermittently due to economic considerations. Similarly, adsorption media can be regenerated through heating, pressure reduction, or chemical washing, and in some cases, desorbed substances may be recovered as valuable byproducts. However, both processes face limitations. Media saturation over time requires periodic regeneration or replacement, which may involve chemical use and generate secondary waste. Selectivity may decline in complex wastewaters containing competing ions or organic matter, and some adsorbents exhibit slow kinetics, necessitating enhancements like particle size reduction or hybridization with processes such as nanofiltration. In ion exchange systems, high influent TSS can clog resin beds, and residual organics may cause resin fouling, reducing efficiency. These challenges can be addressed through appropriate pre-treatment, prefiltration, or scavenger resins.

References

1. Koul, B.; Yadav, D.; Singh, S.; Kumar, M.; Song, M. Insights into the Domestic Wastewater Treatment (DWWT) Regimes: A Review. *Water* **2022**, *14* (21), 3542. https://doi.org/10.3390/w14213542.
2. Shaikh, I. N.; Ahammed, M. M. Granular Media Filtration for On-Site Treatment of Greywater: A Review. *Water Sci. Technol.* **2022**, *86* (5), 992–1016. https://doi.org/10.2166/wst.2022.269.
3. Vigneswaran, S.; Sundaravadivel, M. Recycle and Reuse of Domestic Wastewater. *Wastewater Recycle Reuse Reclam.* **2004**, *1*.
4. Metcalf & Eddy; Abu-Orf, M.; Bowden, G.; Burton, F. L.; Pfrang, W.; Stensel, H. D.; Tchobanoglous, G.; Tsuchihashi, R.; AECOM (Firm). *Wastewater Engineering: Treatment and Resource Recovery*; McGraw Hill Education, 2014.
5. Ranjan, P.; Prem, M. Schmutzdecke-a Filtration Layer of Slow Sand Filter. *Int. J. Curr. Microbiol. Appl. Sci.* **2018**, *7* (07), 637–645.
6. Hube, S.; Eskafi, M.; Hrafnkelsdóttir, K. F.; Bjarnadóttir, B.; Bjarnadóttir, M. Á.; Axelsdóttir, S.; Wu, B. Direct Membrane Filtration for Wastewater Treatment and Resource Recovery: A Review. *Sci. Total Environ.* **2020**, *710*, 136375.
7. Wang, H.; Zhou, H. Understand the Basics of Membrane Filtration. *Chem. Eng. Prog.* **2013**, *109* (4), 33–40.
8. Wang, Y.-N.; Goh, K.; Li, X.; Setiawan, L.; Wang, R. Membranes and Processes for Forward Osmosis-Based Desalination: Recent Advances and Future Prospects. *Desalination* **2018**, *434*, 81–99.

9. Wei, J.; Qiu, C.; Tang, C. Y.; Wang, R.; Fane, A. G. Synthesis and Characterization of Flat-Sheet Thin Film Composite Forward Osmosis Membranes. *J. Membr. Sci.* **2011**, *372* (1–2), 292–302.
10. Xia, L.; Andersen, M. F.; Hélix-Nielsen, C.; McCutcheon, J. R. Novel Commercial Aquaporin Flat-Sheet Membrane for Forward Osmosis. *Ind. Eng. Chem. Res.* **2017**, *56* (41), 11919–11925.
11. Arena, J. T.; Manickam, S. S.; Reimund, K. K.; Brodskiy, P.; McCutcheon, J. R. Characterization and Performance Relationships for a Commercial Thin Film Composite Membrane in Forward Osmosis Desalination and Pressure Retarded Osmosis. *Ind. Eng. Chem. Res.* **2015**, *54* (45), 11393–11403.
12. Ren, J.; McCutcheon, J. R. A New Commercial Thin Film Composite Membrane for Forward Osmosis. *Desalination* **2014**, *343*, 187–193.
13. Zhang, S.; Wang, K. Y.; Chung, T.-S.; Jean, Y.; Chen, H. Molecular Design of the Cellulose Ester-Based Forward Osmosis Membranes for Desalination. *Chem. Eng. Sci.* **2011**, *66* (9), 2008–2018.
14. Wang, R.; Shi, L.; Tang, C. Y.; Chou, S.; Qiu, C.; Fane, A. G. Characterization of Novel Forward Osmosis Hollow Fiber Membranes. *J. Membr. Sci.* **2010**, *355* (1–2), 158–167.
15. Setiawan, L.; Wang, R.; Shi, L.; Li, K.; Fane, A. G. Novel Dual-Layer Hollow Fiber Membranes Applied for Forward Osmosis Process. *J. Membr. Sci.* **2012**, *421*, 238–246.
16. Wei, J.; Qiu, C.; Wang, Y.-N.; Wang, R.; Tang, C. Y. Comparison of NF-like and RO-like Thin Film Composite Osmotically Driven Membranes—Implications for Membrane Selection and Process Optimization. *J. Membr. Sci.* **2013**, *427*, 460–471.
17. Qi, S.; Li, Y.; Zhao, Y.; Li, W.; Tang, C. Y. Highly Efficient Forward Osmosis Based on Porous Membranes Applications and Implications. *Environ. Sci. Technol.* **2015**, *49* (7), 4690–4695.
18. Findley, M. E. Vaporization through Porous Membranes. *Ind. Eng. Chem. Process Des. Dev.* **1967**, *6* (2), 226–230.
19. Khayet, M. Membranes and Theoretical Modeling of Membrane Distillation: A Review. *Adv. Colloid Interface Sci.* **2011**, *164* (1–2), 56–88.
20. Aptel, P.; Challard, N.; Cuny, J.; Neel, J. Application of the Pervaporation Process to Separate Azeotropic Mixtures. *J. Membr. Sci.* **1976**, *1*, 271–287.
21. Ong, Y. K.; Shi, G. M.; Le, N. L.; Tang, Y. P.; Zuo, J.; Nunes, S. P.; Chung, T.-S. Recent Membrane Development for Pervaporation Processes. *Prog. Polym. Sci.* **2016**, *57*, 1–31.
22. Shaffer, D. L.; Arias Chavez, L. H.; Ben-Sasson, M.; Romero-Vargas Castrillón, S.; Yip, N. Y.; Elimelech, M. Desalination and Reuse of High-Salinity Shale Gas Produced Water: Drivers, Technologies, and Future Directions. *Environ. Sci. Technol.* **2013**, *47* (17), 9569–9583.
23. Song, Y.; Li, R.; Pan, F.; He, Z.; Yang, H.; Li, Y.; Yang, L.; Wang, M.; Wang, H.; Jiang, Z. Ultrapermeable Graphene Oxide Membranes with Tunable Interlayer Distances via Vein-like Supramolecular Dendrimers. *J. Mater. Chem. A* **2019**, *7* (31), 18642–18652.
24. Deshmukh, A.; Boo, C.; Karanikola, V.; Lin, S.; Straub, A. P.; Tong, T.; Warsinger, D. M.; Elimelech, M. Membrane Distillation at the Water-Energy Nexus: Limits, Opportunities, and Challenges. *Energy Environ. Sci.* **2018**, *11* (5), 1177–1196.
25. Ghaffour, N.; Soukane, S.; Lee, J.-G.; Kim, Y.; Alpatova, A. Membrane Distillation Hybrids for Water Production and Energy Efficiency Enhancement: A Critical Review. *Appl. Energy* **2019**, *254*, 113698.
26. Zhang, R.; Liu, Y.; He, M.; Su, Y.; Zhao, X.; Elimelech, M.; Jiang, Z. Antifouling Membranes for Sustainable Water Purification: Strategies and Mechanisms. *Chem. Soc. Rev.* **2016**, *45* (21), 5888–5924.
27. Sanmartino, J.; Khayet, M.; García-Payo, M.; El-Bakouri, H.; Riaza, A. Treatment of Reverse Osmosis Brine by Direct Contact Membrane Distillation: Chemical Pretreatment Approach. *Desalination* **2017**, *420*, 79–90.

28. Mansour, S.; Giwa, A.; Hasan, S. Novel Graphene Nanoplatelets-Coated Polyethylene Membrane for the Treatment of Reject Brine by Pilot-Scale Direct Contact Membrane Distillation: An Optimization Study. *Desalination* **2018**, *441*, 9–20.
29. Martinetti, C. R.; Childress, A. E.; Cath, T. Y. High Recovery of Concentrated RO Brines Using Forward Osmosis and Membrane Distillation. *J. Membr. Sci.* **2009**, *331* (1–2), 31–39.
30. Chung, H. W.; Swaminathan, J.; Warsinger, D. M. Multistage Vacuum Membrane Distillation (MSVMD) Systems for High Salinity Applications. *J. Membr. Sci.* **2016**, *497*, 128–141.
31. Kim, J.; Kim, J.; Hong, S. Recovery of Water and Minerals from Shale Gas Produced Water by Membrane Distillation Crystallization. *Water Res.* **2018**, *129*, 447–459.
32. Gonzalez, D.; Amigo, J.; Suárez, F. Membrane Distillation: Perspectives for Sustainable and Improved Desalination. *Renew. Sustain. Energy Rev.* **2017**, *80*, 238–259.
33. Wang, T.; Zhang, Y.; Shi, X.; Wu, L.; Zhang, X.; Zhang, S. Polyurethane Hybrid Membranes with Confined Mass Transfer Channels: The Effect of Functionalized Multi-Walled Carbon Nanotubes on Permeation Properties. *Chem. Eng. Sci.* **2019**, *201*, 191–200.
34. Thomas, N.; Mavukkandy, M. O.; Loutatidou, S.; Arafat, H. A. Membrane Distillation Research & Implementation: Lessons from the Past Five Decades. *Sep. Purif. Technol.* **2017**, *189*, 108–127.
35. El-Abbassi, A.; Hafidi, A.; Khayet, M.; García-Payo, M. C. Integrated Direct Contact Membrane Distillation for Olive Mill Wastewater Treatment. *Desalination* **2013**, *323*, 31–38.
36. Zarebska, A.; Nieto, D. R.; Christensen, K. V.; Norddahl, B. Ammonia Recovery from Agricultural Wastes by Membrane Distillation: Fouling Characterization and Mechanism. *Water Res.* **2014**, *56*, 1–10.
37. Damtie, M. M.; Kim, B.; Woo, Y. C.; Choi, J.-S. Membrane Distillation for Industrial Wastewater Treatment: Studying the Effects of Membrane Parameters on the Wetting Performance. *Chemosphere* **2018**, *206*, 793–801.
38. El-Abbassi, A.; Hafidi, A.; García-Payo, M. C.; Khayet, M. Concentration of Olive Mill Wastewater by Membrane Distillation for Polyphenols Recovery. *Desalination* **2009**, *245* (1–3), 670–674.
39. Du, X.; Zhang, Z.; Carlson, K. H.; Lee, J.; Tong, T. Membrane Fouling and Reusability in Membrane Distillation of Shale Oil and Gas Produced Water: Effects of Membrane Surface Wettability. *J. Membr. Sci.* **2018**, *567*, 199–208.
40. Wu, Y.; Kang, Y.; Zhang, L.; Qu, D.; Cheng, X.; Feng, L. Performance and Fouling Mechanism of Direct Contact Membrane Distillation (DCMD) Treating Fermentation Wastewater with High Organic Concentrations. *J. Environ. Sci.* **2018**, *65*, 253–261.
41. Lee, A.; Elam, J. W.; Darling, S. B. Membrane Materials for Water Purification: Design, Development, and Application. *Environ. Sci. Water Res. Technol.* **2016**, *2* (1), 17–42.
42. Baker, R. W. Membrane Technology. *Encycl. Polym. Sci. Technol.* **2002**, *3*.
43. Visakh, P. M.; Nazarenko, O. *Nanostructured Polymer Membranes: Applications*, 1st ed.; Wiley, 2016. https://doi.org/10.1002/9781118831823.
44. Kimura, K.; Amy, G.; Drewes, J. E.; Heberer, T.; Kim, T.-U.; Watanabe, Y. Rejection of Organic Micropollutants (Disinfection by-Products, Endocrine Disrupting Compounds, and Pharmaceutically Active Compounds) by NF/RO Membranes. *J. Membr. Sci.* **2003**, *227* (1–2), 113–121.
45. Urtiaga, A.; Pérez, G.; Ibáñez, R.; Ortiz, I. Removal of Pharmaceuticals from a WWTP Secondary Effluent by Ultrafiltration/Reverse Osmosis Followed by Electrochemical Oxidation of the RO Concentrate. *Desalination* **2013**, *331*, 26–34.
46. Dong, L.; Huang, X.; Wang, Z.; Yang, Z.; Wang, X.; Tang, C. Y. A Thin-Film Nanocomposite Nanofiltration Membrane Prepared on a Support with in Situ Embedded Zeolite Nanoparticles. *Sep. Purif. Technol.* **2016**, *166*, 230–239.

47. Wang, Z.; Zhang, B.; Fang, C.; Liu, Z.; Fang, J.; Zhu, L. Macroporous Membranes Doped with Micro-Mesoporous β-Cyclodextrin Polymers for Ultrafast Removal of Organic Micropollutants from Water. *Carbohydr. Polym.* **2019**, *222*, 114970.
48. Safarpour, M.; Vatanpour, V.; Khataee, A. Preparation and Characterization of Graphene Oxide/TiO2 Blended PES Nanofiltration Membrane with Improved Antifouling and Separation Performance. *Desalination* **2016**, *393*, 65–78.
49. Vinodhini, P. A.; Sudha, P. Removal of Heavy Metal Chromium from Tannery Effluent Using Ultrafiltration Membrane. *Text. Cloth. Sustain.* **2016**, *2* (1), 5.
50. Ghaemi, N.; Madaeni, S. S.; Daraei, P.; Rajabi, H.; Zinadini, S.; Alizadeh, A.; Heydari, R.; Beygzadeh, M.; Ghouzivand, S. Polyethersulfone Membrane Enhanced with Iron Oxide Nanoparticles for Copper Removal from Water: Application of New Functionalized Fe3O4 Nanoparticles. *Chem. Eng. J.* **2015**, *263*, 101–112.
51. Klatt, M.; Beyer, F.; Einfeldt, J. Hospital Wastewater Treatment and the Role of Membrane Filtration–Removal of Micropollutants and Pathogens: A Review. *Water Sci. Technol.* **2022**, *86* (9), 2213–2232.
52. Ottoson, J.; Hansen, A.; Björlenius, B.; Norder, H.; Stenström, T. Removal of Viruses, Parasitic Protozoa and Microbial Indicators in Conventional and Membrane Processes in a Wastewater Pilot Plant. *Water Res.* **2006**, *40* (7), 1449–1457.
53. Gander, M.; Jefferson, B.; Judd, S. Membrane Bioreactors for Use in Small Wastewater Treatment Plants: Membrane Materials and Effluent Quality. *Water Sci. Technol.* **2000**, *41* (1), 205–211.
54. Kiso, Y.; Nishimura, Y.; Kitao, T.; Nishimura, K. Rejection Properties of Non-Phenylic Pesticides with Nanofiltration Membranes. *J. Membr. Sci.* **2000**, *171* (2), 229–237.
55. Glaze, W. H.; Kang, J.-W.; Chapin, D. H. The Chemistry of Water Treatment Processes Involving Ozone, Hydrogen Peroxide and Ultraviolet Radiation. **1987**.
56. Glaze, W. H. Drinking-Water Treatment with Ozone. *Environ. Sci. Technol.* **1987**, *21* (3), 224–230.
57. Deng, Y.; Zhao, R. Advanced Oxidation Processes (AOPs) in Wastewater Treatment. *Curr. Pollut. Rep.* **2015**, *1* (3), 167–176.
58. Oh, W.-D.; Dong, Z.; Lim, T.-T. Generation of Sulfate Radical through Heterogeneous Catalysis for Organic Contaminants Removal: Current Development, Challenges and Prospects. *Appl. Catal. B Environ.* **2016**, *194*, 169–201.
59. Wang, J.; Wang, S. Activation of Persulfate (PS) and Peroxymonosulfate (PMS) and Application for the Degradation of Emerging Contaminants. *Chem. Eng. J.* **2018**, *334*, 1502–1517.
60. Dhamorikar, R. S.; Lade, V. G.; Kewalramani, P. V.; Bindwal, A. B. Review on Integrated Advanced Oxidation Processes for Water and Wastewater Treatment. *J. Ind. Eng. Chem.* **2024**, *138*, 104–122. https://doi.org/10.1016/j.jiec.2024.04.037.
61. Gogate, P. R.; Pandit, A. B. A Review of Imperative Technologies for Wastewater Treatment I: Oxidation Technologies at Ambient Conditions. *Adv. Environ. Res.* **2004**, *8* (3–4), 501–551. https://doi.org/10.1016/S1093-0191(03)00032-7.
62. Beltran, F. J. *Ozone Reaction Kinetics for Water and Wastewater Systems*, 0 ed.; CRC Press, 2003. https://doi.org/10.1201/9780203509173.
63. Völker, J.; Stapf, M.; Miehe, U.; Wagner, M. Systematic Review of Toxicity Removal by Advanced Wastewater Treatment Technologies via Ozonation and Activated Carbon. *Environ. Sci. Technol.* **2019**, *53* (13), 7215–7233. https://doi.org/10.1021/acs.est.9b00570.
64. Babuponnusami, A.; Muthukumar, K. A Review on Fenton and Improvements to the Fenton Process for Wastewater Treatment. *J. Environ. Chem. Eng.* **2014**, *2* (1), 557–572. https://doi.org/10.1016/j.jece.2013.10.011.

65. Zha, F. G.; Yao, D. X.; Hu, Y. B.; Gao, L. M.; Wang, X. M. Integration of US/Fe2+ and Photo-Fenton in Sequencing for Degradation of Landfill Leachate. *Water Sci. Technol.* **2016**, *73* (2), 260–266. https://doi.org/10.2166/wst.2015.487.
66. Agarkoti, C.; Thanekar, P. D.; Gogate, P. R. Cavitation Based Treatment of Industrial Wastewater: A Critical Review Focusing on Mechanisms, Design Aspects, Operating Conditions and Application to Real Effluents. *J. Environ. Manage.* **2021**, *300*, 113786. https://doi.org/10.1016/j.jenvman.2021.113786.
67. Wang, N.; Li, L.; Wang, K.; Huang, X.; Han, Y.; Ma, X.; Wang, M.; Lv, X.; Bai, X. Study and Application Status of Ultrasound in Organic Wastewater Treatment. *Sustainability* **2023**, *15* (21), 15524. https://doi.org/10.3390/su152115524.
68. Martínez-Huitle, C. A.; Brillas, E. Decontamination of Wastewaters Containing Synthetic Organic Dyes by Electrochemical Methods: A General Review. *Appl. Catal. B Environ.* **2009**, *87* (3–4), 105–145. https://doi.org/10.1016/j.apcatb.2008.09.017.
69. Wang, J.; Chen, H. Catalytic Ozonation for Water and Wastewater Treatment: Recent Advances and Perspective. *Sci. Total Environ.* **2020**, *704*, 135249.
70. Bailey, P. S. The Reactions of Ozone with Organic Compounds. *Chem. Rev.* **1958**, *58* (5), 925–1010.
71. Legrini, O.; Oliveros, E.; Braun, A. Photochemical Processes for Water Treatment. *Chem. Rev.* **1993**, *93* (2), 671–698.
72. Alvares, A.; Diaper, C.; Parsons, S. Partial Oxidation by Ozone to Remove Recalcitrance from Wastewaters-a Review. *Environ. Technol.* **2001**, *22* (4), 409–427.
73. Rekhate, C. V.; Shrivastava, J. Decolorization of Azo Dye Solution by Ozone Based Advanced Oxidation Processes: Optimization Using Response Surface Methodology and Neural Network. *Ozone Sci. Eng.* **2020**, *42* (6), 492–506.
74. Sangave, P. C.; Gogate, P. R.; Pandit, A. B. Ultrasound and Ozone Assisted Biological Degradation of Thermally Pretreated and Anaerobically Pretreated Distillery Wastewater. *Chemosphere* **2007**, *68* (1), 42–50.
75. Gottschalk, C.; Libra, J. A.; Saupe, A. *Ozonation of Water and Waste Water: A Practical Guide to Understanding Ozone and Its Applications*; John Wiley & Sons, 2009.
76. Photo-Oxidation Technologies for Advanced Water Treatment. In *Applied Environmental Science and Engineering for a Sustainable Future*; Springer International Publishing: Cham, 2020; pp 221–255. https://doi.org/10.1007/978-3-030-29840-1_10.
77. Drewes, J.; Khan, S. Water Reuse for Drinking Water Augmentation. *Water Qual. Treat.* **2011**, 16–1.
78. Wols, B. A.; Hofman-Caris, C. H. M. Review of Photochemical Reaction Constants of Organic Micropollutants Required for UV Advanced Oxidation Processes in Water. *Water Res.* **2012**, *46* (9), 2815–2827. https://doi.org/10.1016/j.watres.2012.03.036.
79. Andreozzi, R. Advanced Oxidation Processes (AOP) for Water Purification and Recovery. *Catal. Today* **1999**, *53* (1), 51–59. https://doi.org/10.1016/S0920-5861(99)00102-9.
80. Martijn, B. J.; Kruithof, J. C.; Rosenthal, L. P. Design and Implementation of UV/H2O2 Treatment in a Full Scale Drinking Water Treatment Plant. *PWN Technol.* **2007**.
81. Audenaert, W. T. M.; Vermeersch, Y.; Van Hulle, S. W. H.; Dejans, P.; Dumoulin, A.; Nopens, I. Application of a Mechanistic UV/Hydrogen Peroxide Model at Full-Scale: Sensitivity Analysis, Calibration and Performance Evaluation. *Chem. Eng. J.* **2011**, *171* (1), 113–126. https://doi.org/10.1016/j.cej.2011.03.071.
82. Kruithof, J. C.; Kamp, P. C.; Martijn, B. J. UV/H_2 O_2 Treatment: A Practical Solution for Organic Contaminant Control and Primary Disinfection. *Ozone Sci. Eng.* **2007**, *29* (4), 273–280. https://doi.org/10.1080/01919510701459311.

83. Von Sonntag, C. Advanced Oxidation Processes: Mechanistic Aspects. *Water Sci. Technol.* **2008**, *58* (5), 1015–1021. https://doi.org/10.2166/wst.2008.467.
84. Reisz, E.; Schmidt, W.; Schuchmann, H.-P.; Von Sonntag, C. Photolysis of Ozone in Aqueous Solutions in the Presence of Tertiary Butanol. *Environ. Sci. Technol.* **2003**, *37* (9), 1941–1948. https://doi.org/10.1021/es0113100.
85. Miklos, D. B.; Remy, C.; Jekel, M.; Linden, K. G.; Drewes, J. E.; Hübner, U. Evaluation of Advanced Oxidation Processes for Water and Wastewater Treatment – A Critical Review. *Water Res.* **2018**, *139*, 118–131. https://doi.org/10.1016/j.watres.2018.03.042.
86. Watts, M. J.; Linden, K. G. Chlorine Photolysis and Subsequent OH Radical Production during UV Treatment of Chlorinated Water. *Water Res.* **2007**, *41* (13), 2871–2878.
87. Sichel, C.; Garcia, C.; Andre, K. Feasibility Studies: UV/Chlorine Advanced Oxidation Treatment for the Removal of Emerging Contaminants. *Water Res.* **2011**, *45* (19), 6371–6380.
88. Kishimoto, N. State of the Art of UV/Chlorine Advanced Oxidation Processes: Their Mechanism, Byproducts Formation, Process Variation, and Applications. *J. Water Environ. Technol.* **2019**, *17* (5), 302–335. https://doi.org/10.2965/jwet.19-021.
89. Song, K.; Mohseni, M.; Taghipour, F. Application of Ultraviolet Light-Emitting Diodes (UV-LEDs) for Water Disinfection: A Review. *Water Res.* **2016**, *94*, 341–349. https://doi.org/10.1016/j.watres.2016.03.003.
90. Zhang, M.; Dong, H.; Zhao, L.; Wang, D.; Meng, D. A Review on Fenton Process for Organic Wastewater Treatment Based on Optimization Perspective. *Sci. Total Environ.* **2019**, *670*, 110–121. https://doi.org/10.1016/j.scitotenv.2019.03.180.
91. Bari, A. H.; Chawla, A.; Pandit, A. B. Sono-Crystallization Kinetics of K2SO4: Estimation of Nucleation, Growth, Breakage and Agglomeration Kinetics. *Ultrason. Sonochem.* **2017**, *35*, 196–203. https://doi.org/10.1016/j.ultsonch.2016.09.018.
92. Bari, A. H.; Pandit, A. B. Ultrasound-facilitated Particle Breakage: Estimation of Kinetic Parameters Using Population Balance Modelling. *Can. J. Chem. Eng.* **2014**, *92* (12), 2046–2052.
93. Ramisetty, K. A.; Pandit, A. B.; Gogate, P. R. Ultrasound Assisted Preparation of Emulsion of Coconut Oil in Water: Understanding the Effect of Operating Parameters and Comparison of Reactor Designs. *Chem. Eng. Process. Process Intensif.* **2015**, *88*, 70–77. https://doi.org/10.1016/j.cep.2014.12.006.
94. Wei, H.; Rahaman, Md. H.; Zhao, J.; Li, D.; Zhai, J. Hydrogen Peroxide Enhanced Sonophotocatalytic Degradation of Acid Orange 7 in Aqueous Solution: Optimization by B Ox– B Ehnken Design. *J. Chem. Technol. Biotechnol.* **2021**, *96* (9), 2647–2658. https://doi.org/10.1002/jctb.6809.
95. Razavi, T.; Fadaei, A.; Sadeghi, M.; Sedehi, M. Evaluation of the Photosonolysis Process Efficacy for the Removal of Anionic Surfactant Linear Alkyl Benzene Sulfonate from Aqueous Solutions. *J. Ecol. Eng.* **2020**, *21* (6), 1–7. https://doi.org/10.12911/22998993/122192.
96. Raut-Jadhav, S.; Saharan, V. K.; Pinjari, D.; Sonawane, S.; Saini, D.; Pandit, A. Synergetic Effect of Combination of AOP's (Hydrodynamic Cavitation and H2O2) on the Degradation of Neonicotinoid Class of Insecticide. *J. Hazard. Mater.* **2013**, *261*, 139–147. https://doi.org/10.1016/j.jhazmat.2013.07.012.
97. Saharan, V. K.; Badve, M. P.; Pandit, A. B. Degradation of Reactive Red 120 Dye Using Hydrodynamic Cavitation. *Chem. Eng. J.* **2011**, *178*, 100–107. https://doi.org/10.1016/j.cej.2011.10.018.
98. Saharan, V. K.; Rizwani, M. A.; Malani, A. A.; Pandit, A. B. Effect of Geometry of Hydrodynamically Cavitating Device on Degradation of Orange-G. *Ultrason. Sonochem.* **2013**, *20* (1), 345–353. https://doi.org/10.1016/j.ultsonch.2012.08.011.

99. Duan, X.; Yang, S.; Wacławek, S.; Fang, G.; Xiao, R.; Dionysiou, D. D. Limitations and Prospects of Sulfate-Radical Based Advanced Oxidation Processes. *J. Environ. Chem. Eng.* **2020**, *8* (4), 103849. https://doi.org/10.1016/j.jece.2020.103849.
100. Guerra-Rodríguez, S.; Rodríguez, E.; Singh, D.; Rodríguez-Chueca, J. Assessment of Sulfate Radical-Based Advanced Oxidation Processes for Water and Wastewater Treatment: A Review. *Water* **2018**, *10* (12), 1828. https://doi.org/10.3390/w10121828.
101. Ganiyu, S. O.; Van Hullebusch, E. D.; Cretin, M.; Esposito, G.; Oturan, M. A. Coupling of Membrane Filtration and Advanced Oxidation Processes for Removal of Pharmaceutical Residues: A Critical Review. *Sep. Purif. Technol.* **2015**, *156*, 891–914. https://doi.org/10.1016/j.seppur.2015.09.059.
102. Feng, Y.; Yang, L.; Liu, J.; Logan, B. E. Electrochemical Technologies for Wastewater Treatment and Resource Reclamation. *Environ. Sci. Water Res. Technol.* **2016**, *2* (5), 800–831. https://doi.org/10.1039/C5EW00289C.
103. Sudoh, M.; Kodera, T.; Sakai, K.; Zhang, J. Q.; Koide, K. Oxidative Degradation of Aqueous Phenol Effluent with Electrogenerated Fenton's Reagent. *J. Chem. Eng. Jpn.* **1986**, *19* (6), 513–518. https://doi.org/10.1252/jcej.19.513.
104. Garcia-Segura, S.; Garrido, J. A.; Rodríguez, R. M.; Cabot, P. L.; Centellas, F.; Arias, C.; Brillas, E. Mineralization of Flumequine in Acidic Medium by Electro-Fenton and Photoelectro-Fenton Processes. *Water Res.* **2012**, *46* (7), 2067–2076. https://doi.org/10.1016/j.watres.2012.01.019.
105. Chen, W.-S.; Huang, C.-P. Decomposition of Nitrotoluenes in Wastewater by Sonoelectrochemical and Sonoelectro-Fenton Oxidation. *Ultrason. Sonochem.* **2014**, *21* (2), 840–845. https://doi.org/10.1016/j.ultsonch.2013.10.026.
106. Cheng, I. F.; Fernando, Q.; Korte, N. Electrochemical Dechlorination of 4-Chlorophenol to Phenol. *Environ. Sci. Technol.* **1997**, *31* (4), 1074–1078. https://doi.org/10.1021/es960602b.
107. Katsounaros, I.; Kyriacou, G. Influence of Nitrate Concentration on Its Electrochemical Reduction on Tin Cathode: Identification of Reaction Intermediates. *Electrochimica Acta* **2008**, *53* (17), 5477–5484. https://doi.org/10.1016/j.electacta.2008.03.018.
108. Moussa, D. T.; El-Naas, M. H.; Nasser, M.; Al-Marri, M. J. A Comprehensive Review of Electrocoagulation for Water Treatment: Potentials and Challenges. *J. Environ. Manage.* **2017**, *186*, 24–41. https://doi.org/10.1016/j.jenvman.2016.10.032.
109. Crini, G. Non-Conventional Low-Cost Adsorbents for Dye Removal: A Review. *Bioresour. Technol.* **2006**, *97* (9), 1061–1085. https://doi.org/10.1016/j.biortech.2005.05.001.
110. Fu, F.; Wang, Q. Removal of Heavy Metal Ions from Wastewaters: A Review. *J. Environ. Manage.* **2011**, *92* (3), 407–418. https://doi.org/10.1016/j.jenvman.2010.11.011.
111. Perera, M. K.; Englehardt, J. D.; Dvorak, A. C. Technologies for Recovering Nutrients from Wastewater: A Critical Review. *Environ. Eng. Sci.* **2019**, *36* (5), 511–529. https://doi.org/10.1089/ees.2018.0436.
112. Siyal, A. A.; Shamsuddin, M. R.; Low, A.; Rabat, N. E. A Review on Recent Developments in the Adsorption of Surfactants from Wastewater. *J. Environ. Manage.* **2020**, *254*, 109797. https://doi.org/10.1016/j.jenvman.2019.109797.

Sludge Treatment, Disposal, and Resource Recovery

5

Abstract

This chapter provides a comprehensive overview of wastewater sludge management. It coveres composition of sludge, its treatment objectives, processing methods, and strategies for disposal and resource recovery. Sludge contains up to 99.5% water along with organic matter, nutrients, and potential contaminants. Based on source from wastewater treatment plant, it is classified into primary, secondary, mixed, and digested types. Treatment aims to reduce volume, stabilize organic matter, destroy pathogens, recover resources, and ensure regulatory compliance. Key methods include preliminary operations, thickening, stabilization, conditioning, dewatering, and drying. Pretreatment technologiee which can be employed are thermal, ultrasonic, and emerging methods enhance biodegradability, dewaterability, and energy yield. Disposal and reuse options range from landfilling and incineration, use in construction materials, fertilizers, and energy recovery through processes like pyrolysis, gasification, and anaerobic digestion. The chapter highlights a shift toward circular economy approaches, integrating sustainable practices to convert sludge from a waste into a valuable resource.

Wastewater sludge, commonly referred to as biosolids in its treated form, is a by-product generated during the treatment of municipal and industrial wastewater. It appears in solid, semi-solid, or slurry form and is composed mainly of water up to 99.5% alongside a variety of solids. These solids include inert inorganic materials such as sand and salts, essential nutrients like nitrogen and phosphorus, organic matter (proteins, fats, and carbohydrates), microorganisms, and potentially harmful substances including heavy metals and

A. H. Bari et al., *Wastewater Treatment*, Synthesis Lectures on Sustainable Development,
https://doi.org/10.1007/978-3-032-07585-7_5

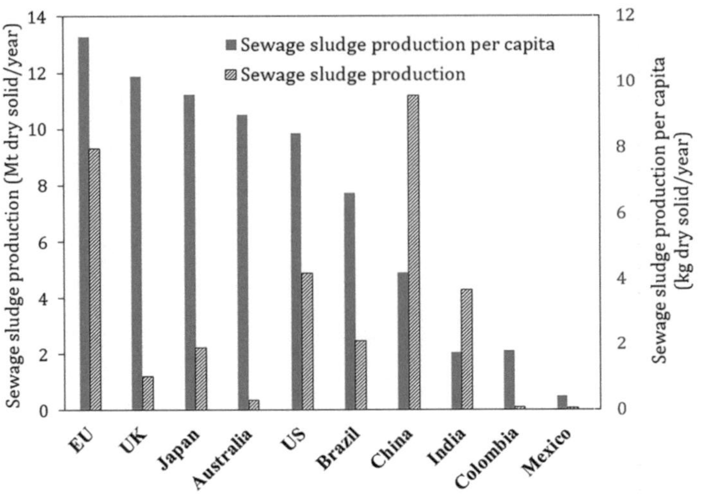

Fig. 5.1 Annual sewage sludge production and per capita sewage sludge production

pathogens [1]. Sludge is generally categorized into primary and secondary types, depending on the treatment stage it originates from [2, 3]. Its composition is influenced by the nature of the incoming influent wastewater and the technologies used during treatment [4].

Due to the increasing urbanization and improvements in wastewater infrastructure, this global sludge production is rising steadily. There are discrepancies in the literature regarding reported values of the total global production of sludge however, the estimated values of sludge production are reported to be as high as 75 Million MT [5]. The contribution to global sludge is not just due to developing countries having significant units of waste water treatment but also due to developing countries like India and China [6]. Approximate estimates for annual sludge production for developed and developing countries are shown in Fig. 5.1. The large amount of waste sludge has raised environmental concerns, particularly with respect to disposal, as sludge occupies significant space and contains both useful and harmful components.

Importantly, sludge also holds potential value. Rich in organic matter and plant nutrients, it can be reused as a biofertilizer in agriculture. Technologies such as anaerobic digestion (AD) are being promoted to convert sludge into biogas, a renewable energy source, especially in colder regions where AD efficiency is often low [7]. A recent trend in value-addition of sludge involves using biosolids to produce products like construction materials, fuels, bioproducts (e.g., biosurfactant, biofloculants, bioplastics, enzymes), and even electricity, though these are at varying production stages and face challenges related to process efficiency, yield, and scalability [8]. Treated biosolids can also be used in construction, such as in roads and as a cement substitute [9]. For small island developing states, where land availability is limited, sustainable management of sludge is especially challenging. However, by focusing on approaches like biogas production, land application of treated sludge, and its use in construction, these regions can reduce environmental risks and promote resource recovery. Ultimately, integrating physical, chemical, and biological treatment methods is key to reducing pollutants in sludge and ensuring compliance with environmental regulations.

5.1 Composition and Characteristics of Sludge

Sludge produced during wastewater treatment can be classified into different types based on the stage of treatment at which it is generated:

(i) Primary Sludge: Obtained from primary clarifiers, this sludge is rich in organic matter and is highly putrescible due to the presence of settleable solids.
(ii) Secondary or Biological Sludge: Formed during biological treatment processes such as the activated sludge process, it primarily consists of microbial biomass and has a high water content.
(iii) Mixed Sludge: A combination of primary and secondary sludge, typically blended before further processing or stabilization to ensure uniformity in treatment.
(iv) Digested Sludge: This is the stabilized form of sludge resulting from aerobic or anaerobic digestion. It is less odorous, more stable, and safer for handling or disposal.

Each type of sludge has distinct characteristics that influence the choice of treatment, stabilization, and disposal or reuse options. However, all types of wastewater sludge primarily consist of water, which accounts for more than 95% of its total mass. Alongside this high moisture content, the sludge contains a significant proportion of organic matter, including proteins, carbohydrates, and lipids. These organic components are often biodegradable but may emit unpleasant odors unless the sludge is properly stabilized. In addition to organics, sludge contains inorganic solids such as sand, grit, and metals that contribute to its overall mass. Essential nutrients like nitrogen, phosphorus, and potassium are also present and, if properly recovered, can be beneficially reused, particularly in agriculture. Table 5.1 summarizes the characteristics of primary and secondary sludge.

The sludge may also harbor various pathogens, including bacteria, viruses, and parasites—which pose health risks and must be addressed through adequate treatment. Furthermore, it may contain contaminants such as heavy metals (e.g., lead and cadmium) [10] that limit the potential for safe reuse as landfill and fertilizers. The selection of appropriate sludge treatment methods depends largely on its physical and chemical properties, including total solids content, volatile solids percentage, and pH. In general, wastewater sludge is a complex mixture dominated by microorganisms and non-degraded organic matter, but it may also contain harmful substances such as pathogens and toxic metals.

Table 5.1 Typical characteristics of primary and digested primary sludge

Parameter	Untreated primary sludge	Untreated secondary sludge
Total solids (TS, % dry solids)	1–6	0.4–1.2
Volatile solids (% of TS)	60–85	60–85
Grease & Fats (% of TS)	5–8	5–12
Protein (% of TS)	20–30	32–41
Nitrogen (N, % of TS)	1.5–4	2.4–5
Phosphorus (P_2O_5, % of TS)	0.8–2.8	2.8–11
Potash (K_2O, % of TS)	0–1	0.5–0.7
Cellulose (% of TS)	8–15	–
Iron (not as sulfide, % of TS)	2–4	–
Silica (SiO_2, % of TS)	15–20	–
pH	5.0–8.0	6.5–8.0
Alkalinity (mg/L as $CaCO_3$)	500–1500	580–1100
Organic acids (mg/L as acetic acid)	200–2000	1100–1700
Energy content (kJ/kg volatile suspended solids)	23,000–29,000	19,000–23,000

5.2 Objectives of Sludge Treatment

The need for effective sludge treatment stems from both environmental and operational imperatives. First, untreated or inadequately treated sludge poses significant risks to public health and the environment due to the presence of pathogenic organisms, toxic heavy metals, and persistent organic pollutants. Direct discharge can contaminate surface and groundwater, release offensive odors, and facilitate disease transmission. Additionally, untreated sludge is bulky and challenging to manage, making treatment essential for volume reduction and safe handling. Therefore, it is necessary to stabilize organic content, reduce pathogens, remove excess water, and convert waste from sludge into a manageable and potentially beneficial product. EPA (2000) [11] further emphasizes that sludge treatment supports resource recovery by enabling the extraction of valuable nutrients like phosphorus and energy sources such as biogas, contributing to sustainability goals.

The key objectives of sludge treatment, therefore, include:

(i) Volume Reduction: Through thickening and dewatering, sludge volume can be reduced by over 80%, making storage, handling, and transportation more economical.

(ii) Stabilization of Organic Matter: Preventing decomposition reduces foul odors, vector attraction, and greenhouse gas emissions, while making the sludge safer and easier to manage.
(iii) Pathogen Reduction: Minimizing the microbial load ensures biosolids are suitable for land application and compliant with public health standards.
(iv) Resource Recovery: Sludge contains nitrogen, phosphorus, and organic carbon, which can be reclaimed for agriculture, energy production, and material recovery.
(v) Environmental Compliance: Treatment ensures that disposal meets regulatory requirements for air, water, and soil protection.

Modern sludge management increasingly emphasizes sustainability and value addition. Reusing biosolids in agriculture, recovering energy through anaerobic digestion, and producing bioplastics, biosurfactants, and construction materials reflect a shift toward circular economy principles [8, 12].

In conclusion, sludge treatment is not only a technical necessity but also a strategic component of sustainable waste management, ensuring environmental protection, public safety, and resource optimization.

5.3 Sludge Treatment Methods

A variety of physical, chemical, and biological procedures are used in sludge treatment in order to achieve the previously mentioned goals (Fig. 5.2). Although sludge comprise of only 1% to 2% of wastewater, sludge treatment and disposal supposes between 50 and 60% of operating costs of wastewater treatment plants [13].

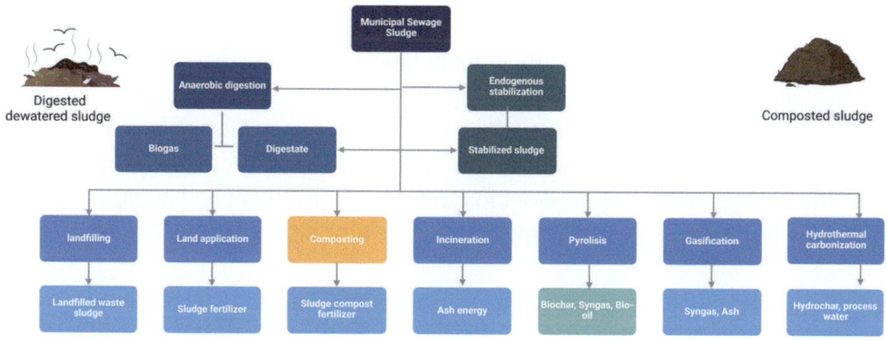

Fig. 5.2 Typical wastewater sludge treatment process

5.3.1 Preliminary Operations

In order to condition the raw sludge for the subsequent treatment stages, the first stage of sludge management involves a number of preparatory procedures. These fundamental processes are mostly mechanical with the aim of homogenizing the sludge and safeguarding the downstream machinery. Usually, three steps are carried out in preliminary operations viz., screening, grinding and blending [1]:

(i) Screening: It is generally the first and it involves physically removing big, inorganic solid objects that have entered the wastewater stream. Materials that could clog pipelines, jam or damage pumps, and hinder the effectiveness of subsequent processes include sticks, rags, plastics, and other debris.
(ii) Grinding: The sludge goes through grinding or comminution after screening. The retained coarse solids particles in sludge are cut and shred into smaller, more uniform particles using specialized machinery. Comminution produces a sludge with a more uniform and fluid texture by decreasing the particle size, making it simpler to pump and mix.
(iii) Blending: It is the last preliminary step, which involves thoroughly mixing sludges from various treatment plant sources, such as the primary and secondary clarifiers, in sizable tanks or basins. Because the properties of sludge can differ greatly depending on where they come from, this step is essential. In order to maintain stable, predictable, and effective performance in the subsequent process of thickening, digestion, and dewatering stages; blending guarantees that the feed entering the following treatment units has a uniform composition, flow rate, and concentration.

5.3.2 Thickening

Thickening is a cost-effective method to reduce the water content of sludge, which typically contains over 95% water, posing challenges in storage, transport, and treatment. By increasing solids concentration, thickening can reduce sludge volume by up to 80%, significantly lowering subsequent treatment costs. Common methods include gravity thickening, dissolved-air flotation (DAF), centrifugation though, several other technologies like separation by hydrocyclones can also be employed [14].

(i) Gravity thickening: It is one of the most popular methods which involves feeding sludge into a large circular tank with a conical bottom. Under the influence of gravity, the solid particles are permitted to settle out of suspension, creating a denser sludge blanket at the bottom that is subsequently gathered. The main disadvantage of gravity thickening is the necessity of large space or area to accommodate capacity needs and allow sufficient settling time.

(ii) DAF: It is another popular technique that works especially well with lighter, less dense sludges. In DAF, millions of tiny air bubbles are created when high pressure dissolves air into a portion of the sludge and is then released in the flotation tank. The solid particles rise to the surface as a result of these bubbles attaching to them, increasing their buoyancy and creating a thicker layer of sludge that can be skimmed off. The requirement of large space, consumption of high energy and higher capital cost of equipment are drawbacks for this method.

(iii) Centrifugation: In centrifugation, high-speed rotational forces using a centrifuge are employed which speeds up the process of separating the denser solid particles from the liquid phase. However, these contain electromechanical parts, e.g. motor and gears, and therefore have higher maintenance costs and are more susceptible to faults.

5.3.3 Stabilization

Sludge stabilization is an essential step in the sludge treatment process. Sludge stabilization is the process through which sewage sludge is biologically decomposed, primarily via aerobic or anaerobic digestion, to a point where its labile (easily decomposable) organic components are transformed into humified material [15]. The untreated sludge generated during primary and secondary treatment stages is rich in biodegradable organic matter, pathogenic microorganisms, and excess moisture. If left unstabilized, this material decomposes rapidly, producing offensive odors, attracting vectors, and posing serious environmental and public health risks. Moreover, land disposal of raw sludge can result in groundwater contamination, greenhouse gas emissions, and pathogen spread. After stabilization, the sludge resists putrefaction, has reduced odor, contains lower levels of pathogenic organisms, and can be disposed of or applied to land without creating environmental or nuisance problems. The final stabilized product, often termed biosolids or digested sludge, is a harmless, humus-like material that can be safely applied to land or further processed [16].

Sludge stabilization can be achieved through various processes like anaerobic digestion, aerobic digestion, and composting [16]. Anaerobic digestion operates in oxygen-free conditions and results in biogas production, while aerobic digestion requires oxygen input and is more suitable for smaller facilities. Composting stabilizes sludge through aerobic microbial activity in combination with carbon-rich bulking agents and is particularly valued for producing a usable organic soil amendment. Chemical stabilization methods, such as lime addition, often lead to increased sludge volume, reduced calorific value and thermal stabilization though effective in pathogen reduction deteriorated effluent quality by returning NH_3 and heavy metals to raw wastewater [17].

5.3.3.1 Anaerobic Digestion

Anaerobic digestion is one of the most mature and widely used biological processes for sludge stabilization. Anaerobic digestion involves the biological breakdown of organic matter in the absence of oxygen, producing biogas (60–70% methane, 30–40% carbon dioxide) and stabilized sludge with significantly lower volatile solids content. Anaerobic digestion achieves almost 40–50% volatile solids reduction, thereby substantially lowering the organic load [18]. Depending on operational needs and environmental conditions, anaerobic digestion can be conducted in two temperature regimes:

(i) Mesophilic digestion which is carried out at around 30–38 °C with retention times of 15–20 days,
(ii) Thermophilic digestion which is carried out at around 50–57 °C with shorter retention times of 10–15 days.

Thermophilic digestion is faster, improved dewatering, higher solid reduction and offers increased pathogen destruction but requires more energy and operational precision.

The process occurs in four main stages: hydrolysis, acidogenesis, acetogenesis, and methanogenesis [18–20].

(i) Hydrolysis: In this stage, complex organic macromolecules such as carbohydrates, proteins, and lipids are broken down into their soluble monomers like sugars, amino acids, and fatty acids by extracellular enzymes. This step is essential for initiating the digestion of particulate matter. A simplified reaction for hydrolysis is:

$$(C-H-O-N)_n + H_2O \rightarrow \text{amino acids} + \text{sugars} + \text{fatty acids}$$

(ii) Acidogenesis: In this stage, the soluble monomers are fermented by acidogenic bacteria into VFAs, alcohols, hydrogen, and carbon dioxide. This phase is also responsible for the generation of acetic acid and ammonia. An example reaction is:

$$C_6H_{12}O_6 \rightarrow 2\,CH_3CH_2OH + 2\,CO_2 \text{ (glucose to ethanol and } CO_2\text{)}$$

Or

$$C_6H_{12}O_6 \rightarrow 2\,CH_3COOH + 2\,H_2 + 2\,CO_2$$

$$\text{(glucose to acetic acid, hydrogen, and } CO_2\text{)}$$

(iii) In the acetogenesis stage, the VFAs produced earlier are converted into acetate, hydrogen, and carbon dioxide. This phase is vital because methanogenic archaea prefer acetate and hydrogen as substrates. A representative reaction is:

$$CH_3CH_2COO^- + 3\,H_2O \rightarrow CH_3COO^- + H^+ + HCO_3^- + 3\,H_2$$

(iv) Finally, methanogenesis involves the conversion of acetic acid and hydrogen into methane by methanogenic archaea under strictly anaerobic conditions. Two key pathways exist:

$$CH_3COOH \rightarrow CH_4 + CO_2 \text{ (acetoclastic methanogenesis)}$$

$$CO_2 + 4\,H_2 \rightarrow CH_4 + 2\,H_2O \text{ (hydrogenotrophic methanogenesis)}$$

The most commonly used reactor configuration for anaerobic digestion is the continuous stirred tank reactor (CSTR), which ensures homogeneous mixing and uniform microbial contact. Other systems include plug flow reactors suitable for high solids input and UASB reactors, which are particularly useful in industrial applications. Standard-rate digesters (also called cold digesters) are relatively simple systems operating without mechanical mixing or heating. These are rarely built nowadays and are typically found in smaller plants due to their simplicity. High-rate digesters improve upon standard systems by including heating, continuous mixing, and uniform feeding, thereby achieving better pathogen reduction and digestion efficiency in significantly shorter times. The tanks can be heated using external heat exchangers and mixed using gas recirculation or mechanical mixers. Two-stage digesters combine a high-rate primary digester with a secondary tank primarily for storage and supernatant separation. Although once popular, they are now less common due to limited performance improvements. A schematic comparison of different anaerobic digester designs is shown in Fig. 5.3. Gas storage and removal components are integrated in all types to capture biogas.

5.3.3.2 Aerobic Digestion

Aerobic digestion is a widely used biological method for stabilizing sewage sludge, particularly in small to medium-sized treatment facilities. The process involves prolonged aeration of sludge in open or closed tanks, allowing aerobic microorganisms to oxidize organic matter into carbon dioxide, water, and microbial biomass. Stabilization is primarily achieved through endogenous respiration, a phase where microorganisms consume their own cellular materials after the depletion of easily biodegradable substrates. Aerobic digesters are simple to operate and cost-effective in terms of capital investment. However, they require continuous aeration, leading to high operational energy costs. The solids retention time (SRT) varies depending on temperature and sludge type is usually 10–20 days, and aerobic systems perform best at temperatures above 15 °C [21]. Notably, unlike anaerobic digestion, aerobic systems do not produce methane and hence do not offer energy recovery. The treated sludge from aerobic digesters is generally odorless, biologically stable, and exhibits improved dewatering characteristics, making it suitable for certain biosolids land applications. Despite energy costs, the process remains an attractive and reliable option due to its operational simplicity and effectiveness in pathogen reduction.

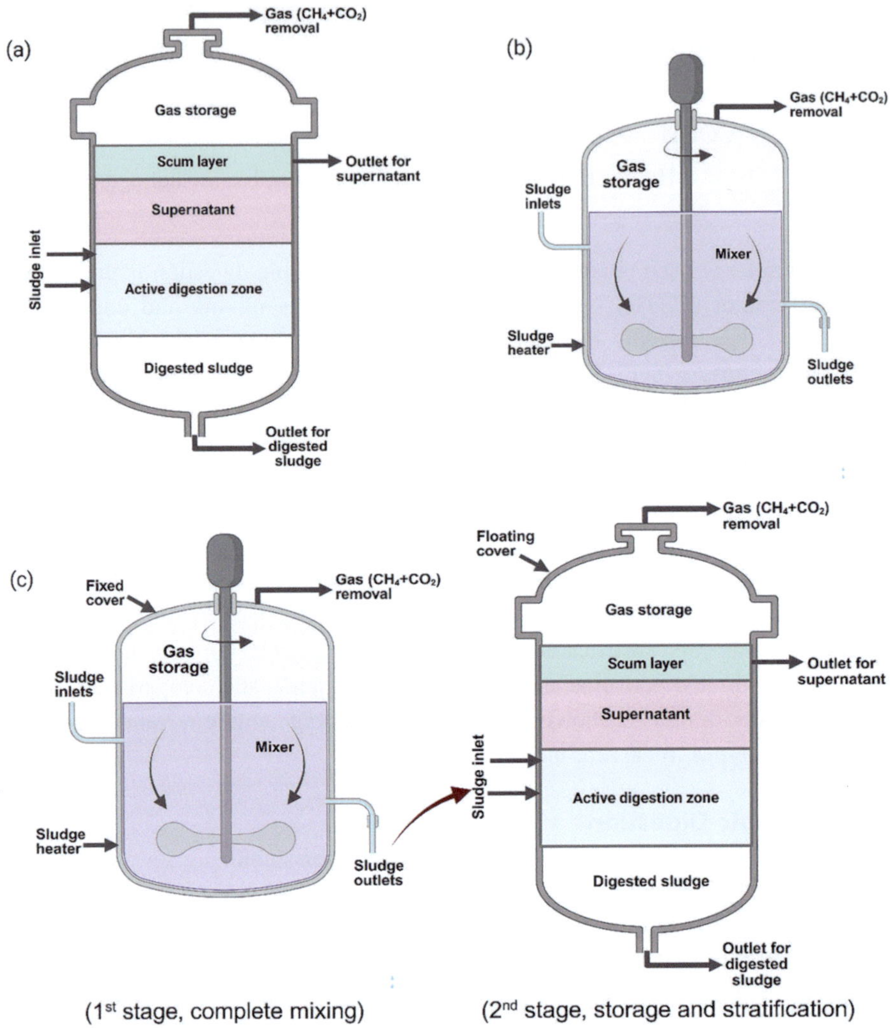

Fig. 5.3 Schematic of different types of Anaerobic digester **a** Standard rate digester **b** High rate digester and **c** Two stage digester

5.3.3.3 Composting

Composting is an effective and increasingly popular biological method for stabilizing sewage sludge, especially in areas with readily available land and carbon-rich bulking materials. It involves the aerobic decomposition of organic matter in the presence of oxygen and elevated temperatures [22]. Dewatered sludge is mixed with bulking agents like sawdust, straw, or wood chips to optimize the carbon-to-nitrogen (C:N) ratio, typically between 26:1 and 30:1. This blend enhances aeration, supports microbial activity, and facilitates the breakdown of organic matter. The end product is a stable, humus-like substance that is safe to work with and can be used as a nutrient-rich soil conditioner on land. The composting process is typically carried out in windrows (large outdoor piles) or in controlled systems like in-vessel composting and aerated static piles. As aerobic microbes degrade the sludge and bulking agents, the process generates heat, raising the temperature above 55 °C. At this temperature, most pathogens, parasites, and weed seeds gets destroyed. To maintain effective composting conditions, periodic turning or aeration is necessary to ensure uniform oxygen distribution and prevent the formation of anaerobic zones that may cause odor issues.

Composting offers several advantages, including effective pathogen reduction, odor control, and the production of a valuable soil amendment. It is particularly well-suited for rural areas where agricultural land is available for biosolids reuse. Co-composting with municipal solid waste or agricultural residues can further improve nutrient balance and enhance the quality of the final product. However, the process requires a relatively large footprint, is sensitive to environmental conditions such as temperature and moisture, and demands careful operational control. Moreover, the finished compost must comply with regulatory standards for stability, heavy metal content, and pathogen levels before it can be safely applied to land [11].

5.3.4 Conditioning

Sewage sludge conditioning is a crucial pretreatment step designed to enhance the dewaterability of sludge before mechanical dewatering operations such as filtration or centrifugation. Its primary objective is to modify the physical and chemical properties of sludge in a way that transforms bound water into free water, thereby allowing easier separation during subsequent dewatering processes. This is achieved by altering the structure of sludge solids and facilitating the formation of larger, more stable flocs, which retain less water. Conditioning methods are broadly categorized into physical and chemical techniques. Traditional chemical conditioning involves the use of inorganic or organic agents such as ferric chloride, lime, calcium oxide, or synthetic organic polymers (e.g., polyelectrolytes). These coagulants and flocculants work by neutralizing the negative surface charges of sludge particles, destabilizing colloids, and facilitating their agglomeration into larger aggregates. Organic polymers, in particular, are advantageous due to their reduced sludge mass addition, cleaner handling, and preservation of the calorific value of dewatered sludge [23]. These polymers are available in various forms like powdered or liquid and must be selected and dosed carefully based on sludge characteristics and the type of dewatering equipment used.

Thermal conditioning is another effective technique where heat is used to rupture microbial cell walls, releasing intracellular water and reducing sludge resistance to filtration. Methods like microwave heating and hydrothermal treatment have demonstrated improvements in dewaterability by decreasing sludge water content and enhancing filtration characteristics. However, these methods may alter the thermal value and organic matter content of sludge due to the incorporation of inorganic residues. Among alternative and more sustainable methods, bioleaching has emerged as an efficient and eco-friendly approach. This microbial-based process utilizes bacteria such as *Acidithiobacillus thiooxidans* and *A. ferrooxidans* to solubilize heavy metals and improve sludge dewatering [24]. Bioleaching not only enhances sludge characteristics but also minimizes the environmental risks associated with metal-laden residues [25]. Overall, conditioning significantly improves sludge dewatering efficiency but also affects nutrient release and organic content. Therefore, selecting a cost-effective, sustainable, and sludge-specific conditioning method is essential for optimizing both treatment performance and environmental outcomes.

5.3.5 Dewatering and Drying

Dewatering which aims to reduce water content up to 20–30% is a crucial unit operation in wastewater sludge management producing a solid-rich "cake" and a liquid stream. This process not only minimizes the volume of sludge, making storage and transportation more economical, but also prepares the material for subsequent treatment or disposal. The liquid stream, particularly from anaerobically digested sludge, is nutrient-rich and usually returned to the treatment plant for sidestream treatment to prevent overloading the main system. There are various technologies used for sludge dewatering, selected based on sludge characteristics, downstream processing requirements, and space availability. Among the most common mechanical dewatering devices are centrifuges, belt filter presses (BFPs), and rotary presses [1]. Centrifuges operate on the principle of high-speed rotation to separate solids from liquids, widely used in both municipal and industrial settings. Belt filter presses, introduced in the 1970s, combine gravity drainage and mechanical pressure to squeeze out water, and they are effective for most types of sludge. Rotary presses, enclosed systems introduced in the 1990s, are known for their safety, odor control, and low noise operation.

After dewatering, sludge drying is carried out to further reduce water content thereby reducing sludge volume and improving handling and disposal. Drying also enhances the sludge's heating value for incineration and reduces leachate generation when landfilled. Among the most commonly used techniques are sludge drying beds, particularly conventional sand drying beds. These beds function through a combination of evaporation and gravity-driven percolation to remove water content. The structure of a drying bed includes multiple layers: coarse gravel at the bottom, finer grit and sand above, and a drainage system of perforated pipes below. Sludge is loaded onto the bed in 200–300 mm thick layers, and water percolates through the medium while evaporating from the surface. Under optimal conditions, this can yield dried sludge cakes with up to 40% solids content within

4–6 weeks. Beds may be open or covered, the latter allowing year-round operation and protection from rain. Drying beds are simple, cost-effective, and require minimal energy or chemical input. However, they demand significant land area and are therefore more suited to rural or suburban settings. In large municipalities or where space is constrained, mechanical alternatives like centrifuges or belt filter presses are preferred for their faster throughput and better operational control. Despite their limitations, drying beds remain a practical and sustainable sludge dewatering method in many regions worldwide.

5.4 Sludge Pretreatment and Enhancement Technologies

Several pretreatment technologies that greatly increase the sustainability and effectiveness of sludge management have been developed as a result of the ongoing evolution of wastewater treatment. These techniques concentrate on increasing energy recovery, improving dewaterability, improving solids reduction, and recovering valuable resources. Most of these techniques are employed after sludge thickening and just before anaerobic digestion. Some of the novel techniques, going beyond traditional methods, provide solutions that can minimize environmental impact, lower operating costs, and add value to waste products.

5.4.1 Thermal Pretreatment

Thermal pretreatment is a widely studied and industrially adopted technique for enhancing the biodegradability and dewaterability of sewage sludge before anaerobic digestion. Various thermal pretreatment processes like CAMBI, BioTHELYS, Exelys, TurboTec, Haarslev, Lysotherm etc., are commercialized with CAMBI being the first one since 1995 [26]. In its fundamental form, thermal pretreatment involves heating sludge to temperatures ranging from 60 °C to 180 °C to destroy the cell walls and makes the proteins accessible for biological degradation [27]. Thermal treatment methods for sludge range from low-temperature heating (<100 °C) for conditioning and dewatering, to middle-temperature ranges (100–200 °C) for hydrolysis or pasteurization, and high-temperature (>200 °C) for pyrolysis and oxidation [28]. The high temperature methods are rather standalone thermal conversion or disposal process and not considered as pretreatment methods.

The low temperature pretreatment at 60° and 80° showed more than 30% increase in gas production both at the laboratory experiment and pilot plant study [29]. However lower temperature requires higher holding time. Middle temperature reduces time and increases dewaterability. Improved dewaterability is reported above 150 °C, though formation of refractory COD compounds was observed at temperatures beyond 180 °C

[27]. A temperature range of 140–160 °C is considered optimal, as it maximizes anaerobic digestion and dewatering performance while minimizing the formation of refractory compounds, which tend to increase sharply above 160 °C potentially deteriorating effluent quality from wastewater treatment plants [30]. In addition, thermal pretreatment reduces odorous compounds and enhances homogenization more effectively than mechanical methods. While energy input is required, the increased biodegradability, improved gas yield, and reduced sludge volume offer a net benefit, making thermal pretreatment a valuable strategy in sustainable sludge management.

5.4.2 Ultrasonic Pretreatment

Ultrasonic disintegration is a well-known method for the break-up particles [31] and microbial cells [32]. The first known study on ultrasound pretreatment of sludge dates back to the early 1990s, with a pioneering work conducted by Tiehm et al. in [33]. Since then, ultrasound pretreatment of sewage sludge has been frontrunner in emerging mechanical techniques aimed at enhancing sludge disintegration and subsequent biogas production. As discussed earlier, ultrasonication causes cavitation which leads to the formation of reactive radicals like hydroxyl (•OH) and hydrogen (•H), coupled with intense hydro-mechanical shear forces, effectively rupturing sludge flocs and microbial cell walls [34]. The dewatering process is aided by this disintegration, which releases bound water from the floc structure and inside the cells. Additionally, ultrasound pretreatment increases the rate and efficiency of anaerobic digestion by increasing the accessibility of the intracellular organic material to microorganisms. Ultrasonic sludge treatment offers compact design, low operational cost, and easy integration into existing systems, while enhancing digester stability, volatile solids destruction, biogas production, and sludge dewater-ability. It also improves biosolids quality, supports denitrification via in situ carbon generation, and helps control filamentous bulking and foaming [35]. On the other hand, ultrasound pretreatment faces key challenge of high capital and operating costs, which currently limit its widespread adoption. Additionally, the lack of long-term performance data from full-scale systems makes design engineers hesitant to recommend it for large-scale implementation.

5.4.3 Emerging Technologies

Emerging technologies for sludge treatment are transforming conventional waste management practices by enhancing sludge disintegration, improving biogas production, and reducing the environmental footprint of treatment plants. These methods include, microwave irradiation, electrochemical oxidation, hydrodynamic cavitation, and various hybrid combinations such as ultrasound combined with alkaline or oxidative treatments etc. Unlike traditional processes, many of these techniques enable better control over sludge properties, enhance volatile solids destruction, and improve dewaterability and pathogen reduction. Table 5.2 summarizes key emerging sludge treatment methods, their mechanisms, benefits and limitations.

These emerging sludge treatment technologies offer significant improvements in process efficiency, resource recovery, and environmental sustainability. Continued research, technological refinement, and cost optimization will be crucial to facilitate their wider adoption in full-scale wastewater treatment systems. Despite challenges related to cost, scalability, and limited full-scale performance data, these technologies hold great potential for future implementation, especially in systems aiming for energy neutrality and circular resource recovery.

5.5 Disposal and Reuse of Bio-Solids

A treated and stabilized sludge often referred as biosolids can either be safely disposed of or repurposed, depending on regulatory requirements, environmental conditions, and site-specific priorities. Traditionally, sludge disposal methods included landfilling, incineration and ocean dumping.

(i) Landfilling: It remains the most widely used method, particularly for dewatered or dried sludge, and must be carried out in properly lined sanitary landfills to prevent leachate leakage into surrounding soil and groundwater.
(ii) Incineration: In cases where biosolids is incinerated, the resulting ash—often containing concentrated heavy metals and other toxic elements—requires careful handling and controlled disposal to avoid secondary environmental impacts.
(iii) Ocean dumping: Once common, is now largely banned due to its irreversible ecological damage.
(iv) Deep well injection: Less common where biosolids are pumped into deep underground geological formations; however, this technique is costly and strictly regulated.

5.5.1 Landfilling

Historically, a significant portion of sewage sludge has been disposed of through landfilling, incineration, or ocean dumping, with only a small share being beneficially reused, particularly in agriculture. Among these methods, landfilling remains a widely used approach due to its operational simplicity and relatively low immediate costs [1]. This process involves transporting dewatered sludge to designated landfill sites where it is buried. Despite its convenience, landfilling is increasingly viewed as a less sustainable option due

Table 5.2 Emerging technologies for sludge treatment

Treatment method	Mechanism involved	Advantages	Disadvantages	References
Microwave Irradiation (MW)	Dipole rotation and ionic conduction produce thermal and athermal effects leading to cell disruption	Rapid heating, enhances solubilization, pathogen reduction, boosts methane yield	High energy input, potential for uneven heating	[36, 37]
Hydrodynamic cavitation	Cavitation from pressure variations induces turbulence and shear stress	Low-cost, scalable, effective for solubilization	Complex reactor design, parameter optimization needed	[38]
Electrochemical pretreatment	Electrolysis generates radicals ($OH^{\bullet}$, $SO_4^{\bullet -}$) that oxidize organic matter and rupture cells	Chemical-free, enhanced solubilization	pH shifts, electrode fouling, electricity-intensive	[39, 40]
Mechanical disintegration	Physical shredding, grinding or milling breaks sludge flocs and cells	Simple, scalable, increases surface area	Equipment wear and moderate energy cost	[41]
Extrusion	Shear, pressure and heat during extrusion disintegrates solids releasing organics	Can form fertilizer pellets, increases accessibility of substrate	High operational energy, equipment maintenance	[42]
Alkaline treatment	High pH causes swelling of solids, breaking of recalcitrant lignin and hemicellulose fractions	Facilitates anaerobic digestion, reduced sludge volume	Potential for nutrient leaching	[43]
H_2O_2/O_3	Excess oxygen oxidizes organic matter	Decrease amount of sludge and increases dewater-ability	CO_2 concentration may high	[44]

(continued)

Table 5.2 (continued)

Treatment method	Mechanism involved	Advantages	Disadvantages	References
MW + Alkaline	Combines thermal, dipole rotation, and high pH for synergistic solubilization and disruption	Increase of solubilization rate, boosts disintegration degree and biogas yield (up to 228%), high SCOD/TCOD ratio (~0.33)	Requires pH control, energy and chemical costs	[45–47]
MW + H_2O_2/ MW + O_3	Thermal decomposition of H_2O_2 enhances OH• generation for oxidation	Higher solubilization (SCOD up to 10,000 mg/L), better than standalone treatments	H_2O_2 dosage must be controlled to avoid inhibition	[44]
US + Alkaline	Synergistic mechanical and chemical disruption enhances protein solubilization	Methane yield increase (~42.1%), improved biodegradability, Increase COD solubilization	Energy and chemical cost, careful pH adjustment needed	[48–51]
US + H_2O_2	Cavitation enhances H_2O_2 oxidation, increases radical formation and solubilization	High soluble COD and shorter time	Requires optimization of pH, H_2O_2 dose, and energy intensive	[52]

to environmental and logistical concerns. One of the primary environmental issues associated with landfilling is the generation of leachate, a highly contaminated liquid that can carry nitrates, heavy metals, organic pollutants, and pathogens. If not properly managed, leachate can seep into surrounding soil and groundwater, posing serious environmental and public health risks. Additionally, the anaerobic degradation of organic matter in landfills produces gases such as methane, which require proper venting and control to prevent safety hazards and greenhouse gas emissions. The viability of landfilling as a long-term strategy is further challenged by the decreasing availability of landfill space and the growing costs associated with environmental compliance. While environmental impacts can be mitigated through proper site selection, engineered barriers, leachate treatment, and gas management systems, these measures add to operational complexity and cost. From a sustainability perspective, landfilling should be considered only when beneficial reuse of sludge is not feasible due to contamination or poor sludge quality. As many European countries are moving toward reducing landfilling to 50% or less [53], the trend is clear:

sludge disposal strategies are shifting toward more sustainable and resource-recovering alternatives.

5.5.2 Incineration

Incineration is a widely adopted method for the disposal of sewage sludge, particularly when the sludge contains toxic or hazardous substances that make other treatment routes unfeasible [17]. This thermal process significantly reduces sludge volume by up to 90% and effectively destroys organic pollutants, minimizes odors, and improves hygiene, making it a compelling alternative to landfilling or long-term storage [54]. Two primary methods of sludge incineration exist: mono-incineration and co-incineration, with mono-incineration being more prevalent. Technologies such as multiple-hearth and fluidized bed furnaces are commonly used due to their energy efficiency and relatively low emissions. Incineration can also enable partial energy recovery, offering benefits in waste-to-energy frameworks.

However, incineration is not without environmental and social challenges. The process produces ash residues typically 10–30% of the dry solids which must be further managed, usually via landfilling. These ashes often concentrate heavy metals, posing potential risks to soil and groundwater if not properly treated or contained. Bottom ash accounts for the majority of the residue and has been found to leach metals unless stabilized through techniques like carbonation [55]. Innovative reuse strategies, such as producing lightweight aggregates from incinerated ash, are emerging to enhance sustainability. Still, concerns persist around air pollution, odour, and social acceptance, especially in densely populated areas. Thus, while incineration offers high efficiency in volume and toxicity reduction, it requires careful management of emissions and ash residues to be a truly sustainable solution. With the advancement of modern flue gas cleaning systems, incineration plants may operate within very stringent emission standards, ensuring minimal impact on air quality. Moreover, when incinerated sludge ash is smelted and repurposed, the risk of heavy metal leaching into groundwater from landfills may be significantly reduced [17].

5.5.3 Reuse as a Building Material

Using sewage sludge or sludge ash to make building and construction materials is a creative and sustainable way to turn a waste product into a useful resource and keep it out of landfills. Sewage sludge and its incinerated ash contain essential oxides such as SiO_2, Al_2O_3, and CaO, making them suitable for use in cementitious materials. Incineration increases oxide concentration by removing volatile and organic content, while the high specific surface area and amorphous nature of sludge ash enhance its pozzolanic activity.

Both sewage sludge and incinerated sewage sludge ash can be used in construction materials by effectively utilizing them as: (i) raw material inputs for producing eco-cement, bricks, ceramics, and lightweight aggregates via sintering; and (ii) supplementary components in cementitious systems, serving as pozzolanic additives, fine aggregates, or fillers [9]. Sewage sludge can replace up to 15% of raw material in eco-cement production without compromising performance compared to traditional Portland cement [56]. Sludge ash, when used at 10–20%, shows only a minor decrease in workability and strength, and its performance improves when combined with other pozzolanic materials like fly ash or slag [57]. Some studies suggest dried sludge and sludge ash can be mixed by 40% and 50% with conventional raw materials to make bricks [58]. Overall the dried sludge or even sludge ash has a potential as a valuable resource in construction materials but the final products must meet certain performance and safety requirements to be viable for use in construction projects.

5.5.4 Fertilizers

The application of biosolids as fertilizer in agriculture is gaining increasing attention due to its nutrient-rich composition and potential for sustainable waste management. Biosolid is rich in primary macronutrients like nitrogen (N) and phosphorus (P) (Table 5.1), contains secondary macronutrients such as calcium (Ca) and magnesium (Mg), especially in alkaline-stabilized forms. It also provides important micronutrients including iron (Fe), copper (Cu), manganese (Mn), zinc (Zn), boron (B), and molybdenum (Mo) [23], all of which contribute to improved soil fertility and plant growth. Thus, when applied appropriately, biosolids not only can enhance soil fertility but also can contribute to improved soil structure and organic carbon content, supporting long-term soil health and productivity. From an agronomic standpoint, the nutrient profile of biosolids is largely influenced by the nature of the influent wastewater and the treatment processes applied. Several studies have shown that the application of sewage sludge can lead to increased crop yields and improved nutrient availability in soil compared to conventional chemical fertilizers [59].

Although case studies from India and other countries support the efficacy of biosolids in improving plant survival and growth, especially in contaminated soils, these findings are best understood as supplementary evidence rather than the core justification for biosolid use [60–62]. Importantly, the use of biosolids is tightly regulated to protect human health and the environment. While low doses of biosolid improve soil health and plant growth, excessive application can lead to the accumulation of toxic heavy metals in crops, posing serious risks to human health and necessitating careful assessment before land use [63]. Sludge must be adequately treated to reduce pathogens and tested for contaminants like heavy metals before its biosolids can be applied to land. Regulations such as those by the U.S. Environmental Protection Agency [11] ensure safe reuse of biosolids in agriculture.

When these guidelines are followed, biosolids serve as a valuable resource for recycling nutrients, improving soil quality, and supporting sustainable agriculture.

5.5.5 Recovery of Energy

Dry sludge primarily contains nontoxic organic compounds, which can be used to recover energy in various forms. There are almost nine options by which energy can be recovered from sewage sludge of which few deals with dry sludge [64]. These nine options are: (1) anaerobic digestion of sewage sludge, (2) production of biofuels from sewage sludge, (3) direct production of electricity from sewage sludge in microbial fuel cells, (4) incineration of sewage sludge with energy recovery, (5) co-incineration of sewage sludge in coal-fired power plants, (6) gasification and pyrolysis of sewage sludge, (7) use of sludge as an energy and raw material source in the production of Portland cement and building materials, (8) supercritical wet oxidation of sewage sludge, and (9) hydrothermal treatment of sewage sludge. Some of the above methods like Anaerobic digestion, incineration etc. are well developed, however some like Hydrothermal Treatment, Microbial Fuel Cells etc. are still in experimental stage. A comprehensive table comparing different sludge-to-energy recovery methods is shown (Table 5.3).

These energy recovery methods need to be evaluated based on their technical and economic feasibility, environmental sustainability, and social acceptability. Out of the incineration, gasification and pyrolysis, the latter is identified as the most favorable method due to its zero-waste profile, lower greenhouse gas emissions, and potential to generate high-value products like bio-oil, char, and syngas [66]. In the comparative assessment, pyrolysis scored 14/16, outperforming gasification (10/16) and incineration (9/16) on criteria such as greenhouse gases emissions, technology maturity, and legislative support. A recent life cycle assessment showed that pyrolysis outperforms both incineration and gasification in terms of climate change impact, particularly when energy systems become less reliant on fossil fuels. Pyrolysis technologies demonstrated the greatest potential for carbon savings, especially when char is used to replace activated carbon or applied to soil as a carbon sink [65]. However, continued development and pilot-scale validation are essential to scale up these energy recovery technologies and integrate them effectively into sustainable sludge management systems.

5.6 Concluding Remarks

While modern waste management practices increasingly emphasize resource recovery and minimize dependency on disposal, safe and environmentally sound disposal remains essential for hazardous, non-recyclable, or highly contaminated sludge streams. A range of options exists, some involve conventional disposal methods, while others focus on

Table 5.3 Comparative analysis of sludge-to-energy recovery methods. Compiled and interpreted from [54, 64–67]

Sr. no	Method	Description	Technology maturity	Energy/ product output	GHG impact	Advantages	Limitations
1	Anaerobic digestion	Biological breakdown of organic matter in absence of oxygen, producing biogas	Mature, widely used	Biogas (~60–70% CH_4), digestate	Low to moderate	Technically simple; high energy recovery; stable digestate	Needs post-treatment; sensitive to temperature/pH
2	Biofuel production (e.g., biodiesel)	Extraction of lipids from sludge for conversion to biodiesel	Emerging	Biodiesel	Medium	Value-added fuel; potential for integration	Fuel yield variable; needs preprocessing
3	Microbial Fuel Cells (MFC)	Electrochemical conversion of organic matter to electricity using microbes	Experimental	Electricity (low power density)	Low	Direct electricity from organics	Low scalability; high cost
4	Incineration (mono/ co-combustion)	Combustion of dried sludge with/without other fuels	Mature, widely used	Heat, electricity, ash	High	Volume reduction > 90%; moderate energy recovery	Air pollution; toxic ash
5	Co-incineration in coal plants	Mixing dried sludge with coal in power plants	Mature in some regions	Heat, electricity	Moderate	Utilizes existing infrastructure	Air pollution, Variable ash quality
6	Gasification	Thermal decomposition in limited oxygen to produce syngas	Developing	Syngas, char	Medium	Higher energy efficiency than incineration	Tar management and disposal; complex operation
7	Pyrolysis	Thermal treatment in absence of oxygen	Emerging to developing	Bio-oil, char, syngas	Low	High-value products; low emissions	Product instability; scale issues

(continued)

Table 5.3 (continued)

Sr. no	Method	Description	Technology maturity	Energy/ product output	GHG impact	Advantages	Limitations
8	Supercritical Water Oxidation (SCWO)	Oxidation of sludge in water above critical point	Experimental/ Pilot	Heat, clean water	Low	Treats toxic sludge; minimal emissions	High pressure and cost systems
9	Hydrothermal treatment	Includes HTL (liquefaction) and HTC (carbonization) to convert sludge under heat/ pressure	Early-stage to developing	Bio-crude, hydrochar	Low	Works with wet sludge; potential fuel	Technology complexity; High energy input

innovative reuse strategies that convert waste into valuable resources. Environmental targets, economic considerations, and compliance with local policies influence the selection of a suitable approach. Disposal becomes necessary when reuse is not feasible due to contamination, poor sludge quality, or limited infrastructure. Current trends indicate a shift from linear disposal models toward circular solutions, where treated sludge is increasingly recognized for its potential applications in construction materials, renewable energy generation, and soil amendment. This transition supports resource recovery and aligns with broader goals in sustainable infrastructure development and integrated waste management.

References

1. Metcalf & Eddy; Abu-Orf, M.; Bowden, G.; Burton, F. L.; Pfrang, W.; Stensel, H. D.; Tchobanoglous, G.; Tsuchihashi, R.; AECOM (Firm). *Wastewater Engineering: Treatment and Resource Recovery*; McGraw Hill Education, 2014.
2. Kumar, V.; Chopra, A.; Kumar, A. A Review on Sewage Sludge (Biosolids) a Resource for Sustainable Agriculture. *Arch. Agric. Environ. Sci.* **2017**, *2* (4), 340–347.
3. Woo, D. C. Y.; Goh, Q. H.; Poh, P. E.; Chew, I. M. L. A Technoeconomic Analysis of Sewage Sludge Valorization for Carbon Emission Reduction. *Biomass Convers. Biorefinery* **2023**, *13* (15), 13591–13604.
4. Fijalkowski, K.; Rorat, A.; Grobelak, A.; Kacprzak, M. J. The Presence of Contaminations in Sewage Sludge–The Current Situation. *J. Environ. Manage.* **2017**, *203*, 1126–1136.
5. Kominko, H.; Gorazda, K.; Wzorek, Z. Sewage Sludge: A Review of Its Risks and Circular Raw Material Potential. *J. Water Process Eng.* **2024**, *63*, 105522.
6. Feng, J.; Burke, I. T.; Chen, X.; Stewart, D. I. Assessing Metal Contamination and Speciation in Sewage Sludge: Implications for Soil Application and Environmental Risk. *Rev. Environ. Sci. Biotechnol.* **2023**, *22* (4), 1037–1058.
7. Alengebawy, A.; Ran, Y.; Osman, A. I.; Jin, K.; Samer, M.; Ai, P. Anaerobic Digestion of Agricultural Waste for Biogas Production and Sustainable Bioenergy Recovery: A Review. *Environ. Chem. Lett.* **2024**, *22* (6), 2641–2668.
8. Tyagi, R.; Surampalli, R. Y.; Yan, S.; Zhang, T. C.; Kao, C.; Lohani, B. Sustainable Sludge Management. *Prod. Value Added Prod. Rest. Va Am.* **2009**.
9. Chang, Z.; Long, G.; Zhou, J. L.; Ma, C. Valorization of Sewage Sludge in the Fabrication of Construction and Building Materials: A Review. *Resour. Conserv. Recycl.* **2020**, *154*, 104606.
10. Alvarez, E. A.; Mochón, M. C.; Sánchez, J. J.; Rodríguez, M. T. Heavy Metal Extractable Forms in Sludge from Wastewater Treatment Plants. *Chemosphere* **2002**, *47* (7), 765–775.
11. U.S. Environmental Protection Agency. *Environmental Regulations and Technology: Control of Pathogens and Vector Attraction in Sewage Sludge*; EPA/625/R-92/013; U.S. Environmental Protection Agency: Washington, DC, 2000.
12. Regitano, J. B.; Rodrigues, M. M.; Martins, G. L.; Osti, J. F.; Viana, D. G.; de Souza, A. J. Sewage Sludge Management for Environmental Sustainability: An Introduction. In *Sustainable Management and Utilization of Sewage Sludge*; Springer, 2022; pp 1–28.
13. Campos, J.; Otero, L.; Franco, A.; Mosquera-Corral, A.; Roca, E. Ozonation Strategies to Reduce Sludge Production of a Seafood Industry WWTP. *Bioresour. Technol.* **2009**, *100* (3), 1069–1073.

14. Senfter, T.; Fritsch, L.; Berger, M.; Kofler, T.; Mayerl, C.; Pillei, M.; Kraxner, M. Sludge Thickening in a Wastewater Treatment Plant Using a Modified Hydrocyclone. *Carbon Resour. Convers.* **2021**, *4*, 132–141.
15. Hartenstein, R. Sludge Decomposition and Stabilization. *Science* **1981**, *212* (4496), 743–749.
16. Lee, D.-J.; Tay, J.-H.; Hung, Y.-T.; He, P. J. Introduction to Sludge Treatment. In *Physicochemical Treatment Processes*; Springer, 2005; pp 677–703.
17. Werther, J.; Ogada, T. Sewage Sludge Combustion. *Prog. Energy Combust. Sci.* **1999**, *25* (1), 55–116.
18. Appels, L.; Baeyens, J.; Degrève, J.; Dewil, R. Principles and Potential of the Anaerobic Digestion of Waste-Activated Sludge. *Prog. Energy Combust. Sci.* **2008**, *34* (6), 755–781.
19. Meegoda, J. N.; Li, B.; Patel, K.; Wang, L. B. A Review of the Processes, Parameters, and Optimization of Anaerobic Digestion. *Int. J. Environ. Res. Public. Health* **2018**, *15* (10), 2224.
20. Ahlawat, S.; Nandal, M. Energy Recovery from Anaerobic Digestion of Wastewater. In *Resilience, Response, and Risk in Water Systems: Shifting Management and Natural Forcings Paradigms*; Springer, 2020; pp 369–379.
21. Qasim, S. R. *Wastewater Treatment Plants: Planning, Design, and Operation*; Routledge, 2017.
22. Mandpe, A.; Kumari, S.; Kumar, S. Composting: A Sustainable Route for Processing of Biodegradable Waste in India. In *Organic waste composting through nexus thinking: Practices, policies, and trends*; Springer International Publishing Cham, 2020; pp 39–60.
23. Andreoli, C. V.; Von Sperling, M.; Fernandes, F. *Sludge Treatment and Disposal*; IWA publishing, 2007.
24. Liu, F.; Zhou, L. Enhancing Dewaterability of Municipal Sewage Sludge through the Combined Approaches of Bioleaching and Fenton Reaction. *Acta Sci Circumst* **2009**, *29* (5), 974–979.
25. Praharaj, S.; Maitra, S.; Hossain, A.; Sagar, L.; Yadav, A. N.; Das, U.; Shankar, T.; Pramanick, B.; Gaikwad, D. Bioleaching Approach for Enhancing Sewage Sludge Dewaterability. In *Sustainable management and utilization of sewage sludge*; Springer, 2022; pp 51–69.
26. Williams, T.; Burrowes, P. Thermal Hydrolysis Offerings and Performance. In *Proceedings of the European Biosolids and Organic Resources Conference*; Edinburgh, Scotland, 2016.
27. Neyens, E.; Baeyens, J. A Review of Thermal Sludge Pre-Treatment Processes to Improve Dewaterability. *J. Hazard. Mater.* **2003**, *98* (1–3), 51–67. https://doi.org/10.1016/S0304-3894(02)00320-5.
28. Waseem, M.; Khan, M. U.; Bokhary, A.; Ahring, B. K. Emerging Trends in Sewage Sludge Pretreatment: Enhancing Treatment Efficiency and Sustainable Waste Management. *Clean. Water* **2025**, *3*, 100080. https://doi.org/10.1016/j.clwat.2025.100080.
29. Hiraoka, M.; Takeda, N.; Sakai, S.; Yasuda, A. Highly Efficient Anaerobic Digestion with Thermal Pretreatment. *Water Sci. Technol.* **1985**, *17* (4–5), 529–539. https://doi.org/10.2166/wst.1985.0157.
30. Devos, P.; Haddad, M.; Carrère, H. Thermal Hydrolysis of Municipal Sludge: Finding the Temperature Sweet Spot: A Review. *Waste Biomass Valorization* **2021**, *12* (5), 2187–2205. https://doi.org/10.1007/s12649-020-01130-1.
31. Bari, A. H.; Pandit, A. B. Ultrasound-facilitated Particle Breakage: Estimation of Kinetic Parameters Using Population Balance Modelling. *Can. J. Chem. Eng.* **2014**, *92* (12), 2046–2052.
32. Gogate, P. R.; Pandit, A. B. Application of Cavitational Reactors for Cell Disruption for Recovery of Intracellular Enzymes. *J. Chem. Technol. Biotechnol. Int. Res. Process Environ. Clean Technol.* **2008**, *83* (8), 1083–1093.
33. Tiehm, A.; Nickel, K.; Neis, U. The Use of Ultrasound to Accelerate the Anaerobic Digestion of Sewage Sludge. *Water Sci. Technol.* **1997**, *36* (11), 121–128.
34. Chu, C.; Lee, D.; Chang, B.-V.; You, C.; Tay, J. "Weak" Ultrasonic Pre-Treatment on Anaerobic Digestion of Flocculated Activated Biosolids. *Water Res.* **2002**, *36* (11), 2681–2688.

35. Khanal, S. K.; Grewell, D.; Sung, S.; Van Leeuwen, J. Ultrasound Applications in Wastewater Sludge Pretreatment: A Review. *Crit. Rev. Environ. Sci. Technol.* **2007**, *37* (4), 277–313.
36. Ahn, J.-H.; Shin, S. G.; Hwang, S. Effect of Microwave Irradiation on the Disintegration and Acidogenesis of Municipal Secondary Sludge. *Chem. Eng. J.* **2009**, *153* (1–3), 145–150.
37. Eskicioglu, C.; Terzian, N.; Kennedy, K. J.; Droste, R. L.; Hamoda, M. Athermal Microwave Effects for Enhancing Digestibility of Waste Activated Sludge. *Water Res.* **2007**, *41* (11), 2457–2466.
38. Bhat, A. P.; Gogate, P. R. Cavitation-Based Pre-Treatment of Wastewater and Waste Sludge for Improvement in the Performance of Biological Processes: A Review. *J. Environ. Chem. Eng.* **2021**, *9* (2), 104743.
39. Chen, Y.; Chen, H.; Li, J.; Xiao, L. Rapid and Efficient Activated Sludge Treatment by Electro-Fenton Oxidation. *Water Res.* **2019**, *152*, 181–190.
40. Feki, E.; Battimelli, A.; Sayadi, S.; Dhouib, A.; Khoufi, S. High-Rate Anaerobic Digestion of Waste Activated Sludge by Integration of Electro-Fenton Process. *Molecules* **2020**, *25* (3), 626.
41. Atelge, M.; Atabani, A.; Banu, J. R.; Krisa, D.; Kaya, M.; Eskicioglu, C.; Kumar, G.; Lee, C.; Yildiz, Y.; Unalan, S. A Critical Review of Pretreatment Technologies to Enhance Anaerobic Digestion and Energy Recovery. *Fuel* **2020**, *270*, 117494.
42. Victorin, M.; Davidsson, Å.; Wallberg, O. Characterization of Mechanically Pretreated Wheat Straw for Biogas Production. *Bioenergy Res.* **2020**, *13* (3), 833–844.
43. Elalami, D.; Monlau, F.; Carrere, H.; Abdelouahdi, K.; Oukarroum, A.; Zeroual, Y.; Barakat, A. Effect of Coupling Alkaline Pretreatment and Sewage Sludge Co-Digestion on Methane Production and Fertilizer Potential of Digestate. *Sci. Total Environ.* **2020**, *743*, 140670.
44. Yin, G.; Liao, P. H.; Lo, K. V. An Ozone/Hydrogen Peroxide/Microwave-Enhanced Advanced Oxidation Process for Sewage Sludge Treatment. *J. Environ. Sci. Health Part A* **2007**, *42* (8), 1177–1181.
45. Jang, J.-H.; Ahn, J.-H. Effect of Microwave Pretreatment in Presence of NaOH on Mesophilic Anaerobic Digestion of Thickened Waste Activated Sludge. *Bioresour. Technol.* **2013**, *131*, 437–442.
46. Yang, Q.; Yi, J.; Luo, K.; Jing, X.; Li, X.; Liu, Y.; Zeng, G. Improving Disintegration and Acidification of Waste Activated Sludge by Combined Alkaline and Microwave Pretreatment. *Process Saf. Environ. Prot.* **2013**, *91* (6), 521–526.
47. Chang, C.-J.; Tyagi, V. K.; Lo, S.-L. Effects of Microwave and Alkali Induced Pretreatment on Sludge Solubilization and Subsequent Aerobic Digestion. *Bioresour. Technol.* **2011**, *102* (17), 7633–7640.
48. Wang, F.; Wang, Y.; Ji, M. Mechanisms and Kinetics Models for Ultrasonic Waste Activated Sludge Disintegration. *J. Hazard. Mater.* **2005**, *123* (1–3), 145–150.
49. Liu, X.; Liu, H.; Chen, J.; Du, G.; Chen, J. Enhancement of Solubilization and Acidification of Waste Activated Sludge by Pretreatment. *Waste Manag.* **2008**, *28* (12), 2614–2622.
50. Tian, X.; Wang, C.; Trzcinski, A. P.; Lin, L.; Ng, W. J. Insights on the Solubilization Products after Combined Alkaline and Ultrasonic Pre-Treatment of Sewage Sludge. *J. Environ. Sci.* **2015**, *29*, 97–105.
51. Seng, B.; Khanal, S. K.; Visvanathan, C. Anaerobic Digestion of Waste Activated Sludge Pre-treated by a Combined Ultrasound and Chemical Process. *Environ. Technol.* **2010**, *31* (3), 257–265.
52. Yuan, D.; Zhou, X.; Jin, W.; Han, W.; Chi, H.; Ding, W.; Huang, Y.; He, Z.; Gao, S.; Wang, Q. Effects of the Combined Utilization of Ultrasonic/Hydrogen Peroxide on Excess Sludge Destruction. *Water* **2021**, *13* (3), 266.

53. Collivignarelli, M. C.; Abbà, A.; Frattarola, A.; Carnevale Miino, M.; Padovani, S.; Katsoyiannis, I.; Torretta, V. Legislation for the Reuse of Biosolids on Agricultural Land in Europe: Overview. *Sustainability* **2019**, *11* (21), 6015.
54. Fytili, D.; Zabaniotou, A. Utilization of Sewage Sludge in EU Application of Old and New Methods—A Review. *Renew. Sustain. Energy Rev.* **2008**, *12* (1), 116–140.
55. Arickx, S.; Van Gerven, T.; Vandecasteele, C. Accelerated Carbonation for Treatment of MSWI Bottom Ash. *J. Hazard. Mater.* **2006**, *137* (1), 235–243.
56. Lin, Y.; Zhou, S.; Li, F.; Lin, Y. Utilization of Municipal Sewage Sludge as Additives for the Production of Eco-Cement. *J. Hazard. Mater.* **2012**, *213*, 457–465.
57. Chen, Z.; Poon, C. S. Comparative Studies on the Effects of Sewage Sludge Ash and Fly Ash on Cement Hydration and Properties of Cement Mortars. *Constr. Build. Mater.* **2017**, *154*, 791–803.
58. Tay, J.-H.; Show, K.-Y. Resource Recovery of Sludge as a Building and Construction Material—a Future Trend in Sludge Management. *Water Sci. Technol.* **1997**, *36* (11), 259–266.
59. Epstein, E. *Land Application of Sewage Sludge and Biosolids*; CRC Press, 2002.
60. Chitdeshwari, T.; Savithri, P.; Mahimairaja, S. Effect of Sewage Biosolid Composts on the Yield of Crops. *Indian J. Environ. Prot.* **2001**, *21* (10), 911–912.
61. Singh, R.; Agrawal, M. Variations in Heavy Metal Accumulation, Growth and Yield of Rice Plants Grown at Different Sewage Sludge Amendment Rates. *Ecotoxicol. Environ. Saf.* **2010**, *73* (4), 632–641.
62. Singh, R.; Agrawal, M. Effect of Different Sewage Sludge Applications on Growth and Yield of Vigna Radiata L. Field Crop: Metal Uptake by Plant. *Ecol. Eng.* **2010**, *36* (7), 969–972.
63. Singh, R. P.; Agrawal, M. Potential Benefits and Risks of Land Application of Sewage Sludge. *Waste Manag.* **2008**, *28* (2), 347–358.
64. Rulkens, W. Sewage Sludge as a Biomass Resource for the Production of Energy: Overview and Assessment of the Various Options. *Energy Fuels* **2008**, *22* (1), 9–15.
65. Chang, H.; Yuan, J.; Zhao, Y.; Bisinella, V.; Damgaard, A.; Christensen, T. H. Carbon Footprints of Incineration, Pyrolysis, and Gasification for Sewage Sludge Treatment. *Resour. Conserv. Recycl.* **2025**, *212*, 107939. https://doi.org/10.1016/j.resconrec.2024.107939.
66. Samolada, M. C.; Zabaniotou, A. Comparative Assessment of Municipal Sewage Sludge Incineration, Gasification and Pyrolysis for a Sustainable Sludge-to-Energy Management in Greece. *Waste Manag.* **2014**, *34* (2), 411–420.
67. Salva, J.; Sečkár, M.; Schwarz, M.; Samešová, D.; Mordáčová, M.; Poništ, J.; Veverková, D. Analysis of the Current State of Sewage Sludge Treatment from the Perspective of Current European Directives. *Environ. Sci. Eur.* **2025**, *37* (1), 59. https://doi.org/10.1186/s12302-025-01097-7.

MIX
Papier aus verantwortungsvollen Quellen
Paper from responsible sources
FSC® C105338

If you have any concerns about our products,
you can contact us on
ProductSafety@springernature.com

In case Publisher is established outside the EU,
the EU authorized representative is:
Springer Nature Customer Service Center GmbH
Europaplatz 3, 69115 Heidelberg, Germany

Printed by Libri Plureos GmbH
in Hamburg, Germany